Demographic Research Monographs

A Series of the Max Planck Institute
for Demographic Research

Editor-in-chief

James W. Vaupel
Max Planck Institute for Demographic Research,
Rostock, Germany

For further volumes:
http://www.springer.com/series/5521

Wenke Apt

Germany's New Security Demographics

Military Recruitment in the Era of Population Aging

 Springer

Wenke Apt

ISSN 1613-5520
ISBN 978-94-007-6963-2   ISBN 978-94-007-6964-9 (eBook)
DOI 10.1007/978-94-007-6964-9
Springer Dordrecht Heidelberg New York London

Library of Congress Control Number: 2013952746

Foreword

This volume, *Germany's New Security Demographics: Military Recruitment in the Era of Population Aging* by Dr. Wenke Apt, is the eleventh book of a series of Demographic Research Monographs published by the Springer Verlag. Dr. Apt is currently working as a scientific coordinator and consultant at the department of "Demographic Change and Futures Studies" at VDI/VDE-IT, a project managing agency for the Federal Ministry of Education and Research in Germany. From 2009 to 2011, she worked as a researcher in the "International Security" division of the Stiftung Wissenschaft und Politik (German Institute for International and Security Affairs – SWP). She is a graduate of Duke University and the European Doctoral School of Demography. This book is based on her doctoral dissertation, which she completed at the Max Planck Institute for Demographic Research and defended at the University of Potsdam in 2010. This revised version includes new findings and developments, in particular with regard to the suspension of compulsory military service in Germany in 2011.

The book focuses on the impact of demographic change on military recruitment. The first chapter sets the stage by introducing the research framework of military manpower demand and supply, and by reviewing new demographic constraints to foreign policymaking. Chapter 2 then provides a detailed account of the military environment, including a discussion of the economic, technological, socio-cultural and geostrategic environments, and about the military roles, missions, and organization. In Chap. 3, Dr. Apt considers how demographic change may affect national security, and in Chap. 4, she discusses the types of manpower needed by the military. These chapters are based on a thorough review and thoughtful assessment of the literature. In Chap. 5, the book presents the results of elaborate, multivariate, statistical analyses carried out to study the determinants of military manpower supply. The statistical inference represents the core of the quantitative analysis undertaken by Dr. Apt. This analysis is noteworthy for three reasons: it is apparently the first such analysis of its kind undertaken for Germany; it is carefully done with thoughtful caveats whenever appropriate; and the findings are of considerable interest and value. In Chap. 6, Dr. Apt uses demographic trends to forecast the manpower available for military recruitment in 2030. Finally, in Chap. 7, she summarizes her findings with an emphasis on how demographic change may constrain the range of available foreign policy options and reviews best practices of military recruitment.

This book is impressive in its careful discussion of the wide range of issues that must be considered when evaluating the implications of demographic change on military manpower opportunities and hence on German foreign policy overall. This book is not only a substantial effort of literature review with more than 400 references plus extensive statistical analysis – it is a pioneering endeavor of such review and analysis. There has been no research in any country comparable to this work; there has been no unified analysis of the interrelationships between demographic change and national security. In particular, nothing approaching the scope of this work has been attempted for Germany. The issues that Dr. Apt addresses are of fundamental significance to Germany, and the severe problems that she foresees are critically important.

Given the synoptic scope of this book, the judicious comparison of alternative arguments and perspectives, the careful interpretation of statistical analyses and forecasting models, and the innovative effort to study how demographic change will affect foreign and security policy, this book is an important scientific contribution to interdisciplinary knowledge at the interface of demography and security studies. It draws on theories, data and concepts of demography, sociology, economics and foreign affairs. Its findings will be of interest for scholars and students in these fields – and beyond. Furthermore, her findings will provide useful insights to analysts and decision-makers in the field of foreign policy.

This book is not the final word, as Dr. Apt stresses in her cautious warnings about limitations imposed by lack of data and uncertainties resulting from disagreements among experts. It is, however, an impressive start on the kind of comprehensive, analytical studies that are needed to put the interrelationship of demography and foreign policy on a firmer foundation of fact and knowledge.

Rostock, Germany and Odense, Denmark   James W. Vaupel
June 2013   Editor-in-Chief

Acknowledgements

This book is a selective and updated version of my doctoral thesis entitled "German Foreign and Security Policy in Transition: New Constraints of Demographic Change" successfully defended at the University of Potsdam in June 2010. It has been revised to incorporate the written and oral comments of my advisors.

Since success is never an orphan, I would like to thank a number of people and organizations that were integral to the completion of my research. First and foremost, I am sincerely grateful to my doctoral advisors, for their precious time, intellectual input, and moral support. I thank Erhard Stölting and Gerhard Kümmel for helping me to determine the shape, scope, and contribution of my research. I thank James W. Vaupel for his valuable assistance and guidance over many years, and his seemingly never-ending enthusiasm about opening new areas of research and teaching new ways of looking at things.

I am indebted to Gabriele Doblhammer in many respects and, most notably, thank her for her assistance in finding solutions to empirical challenges throughout my doctoral research. In equal measure, I thank Andreas Edel for his advice and encouragement. I also recognize, and am grateful for, the financial support of the Max Planck Society.

I am much obliged to the German Federal Ministry of Defense, in particular Colonel Hans Hermann Paape and his colleagues at the Personnel, Social Services and Central Affairs Directorate.

For their genial support, critical remarks, and technical assistance, I am grateful to Caroline Berghammer, Angelika Frederking, Katharina Frosch, Jutta Gampe, Esther Geisler, Rainer Heuer, Saskia Hin, Jörg Jacobs, Judith Kelley, Christian Leuprecht, Heiner Maier, Vegard Skirbekk, Matthias Teichner, Marlen Toch, Joris Van Bladel, Dirk Vieregg, Christian Wegner, Jana Wenzel, and Christina Westphal. I also thank Richard Jackson, who initiated my interest in the security implications of demographic change.

This book is dedicated to my family: my brother, who I thank for his unconditional care, and my parents, who always provided me with the confiding guidance of a compass and the sense of security that one only finds in a safe harbor.

Contents

List of Figures

List of Tables

Chapter 1
Introduction

1.1 Relevance

Military recruitment represents one of the most difficult staffing challenges and a prime example of the multidimensionality of labor market decisions. In addition to the quantitative requirements of manpower demand, successful candidates must fulfill a range of qualitative criteria, such as physical abilities and cognitive skills to master complex military systems and operations, high moral standards, age limits, and citizenship requirements. Moreover, successful applicants usually have a strong sense of duty, value teamwork and are willing to take responsibility and face challenging circumstances (NRC 2003: 219). In the future, soldierly requirements will continue to rise or, at least, remain high. The reasons are multifaceted, one being the complexity and sensitivity of military missions, another being the trend towards an increasing substitution of capital for labor, which will further increase the complexity of military technology (Leuprecht 2010: 47). According to Dandeker and Mason (2010: 219), these requirements "place a premium on more intelligent and diverse service personnel who are able to respond flexibly and with initiative, and hence on the recruitment and retention of better educated and qualified personnel." Similarly, Krebs and Levy (2001: 67) argue that advanced armaments require "highly educated and technically proficient manpower to operate and maintain them".

Although one can only speculate about manpower requirements for future military missions, there is general agreement that armed conflicts will continue to be a human endeavor (Kagan 2006: 97; JOE 2008: 6) and that manpower will remain "the principal determinant of a state's military power and essential to success in almost any imaginable future conflict" (Korb et al. 2006). In particular, the contemporary military role set of expeditionary war-fighting, peacekeeping and nation-building displays a strong human element. Albeit technological improvements, there exists no substitute for large ground forces that are capable of accomplishing the critical tasks of judgment, control and cooperation, or socio-political reorganization (Kagan 2006: 106–108). On that effect, Biddle (2002: ix) cautions that "what Afghanistan really shows is that the wars of tomorrow – like those of

W. Apt, *Germany's New Security Demographics: Military Recruitment in the Era of Population Aging*, Demographic Research Monographs, DOI 10.1007/978-94-007-6964-9_1, © Springer Science+Business Media Dordrecht 2014

yesterday – will continue to require skilled, motivated forces on the ground". Similarly, Boot (2005: 104) argues that "defeating terrorism [...] requires putting boots on the ground and engaging in nation building". For the Bundeswehr, as an allied partner, this means a continued high demand for "well-trained, capable and motivated soldiers [...that] are crucial for operational readiness" (Federal Ministry of Defense 2006a: 112).

However, in light of the demographic changes under way, the absolute size and relative share of the youth population will shrink in Germany, as in all other Western industrialized nations. As a result, the private and public sector will be both facing labor market shortages and competing for the same limited pool of qualified personnel (Leuprecht 2010: 47). In view of the military's particular dependence on the constant inflow of young people to fill the junior ranks in the military personnel structure, many organizational measures to counter demographic change in the wider economy, such as the delay of retirement or the increased usage of information and communication technologies, will only have marginal effects and will not solve the ultimate question of "Who will serve?" (Dandeker and Mason 2010: 215). The Bundeswehr already fails to meet its recruiting goals. For several years, 7,000 vacancies have been left unfilled (Federal Ministry of Defense 2010: 25). Mostly affected are the areas of information technology, electronic engineering, medical services, special forces, and aviation (Federal Ministry of Defense 2006b: 46).

In the broader social environment of Western militaries, a range of factors adds to the absolute decline in youth and therefore exacerbates the challenge of recruiting sufficient numbers of new soldiers. These arise from different trends, including the overlap of the military's recruitment target population with that of other major civilian institutions, the emergence of the knowledge economy and increasing levels of participation in higher education, an increasingly heterogeneous population profile within the target recruitment pool, poor health among young people, as well as socio-cultural changes related to the rise of individualism and the acceptable balance between career and family (Dandeker and Mason 2010: 209f.; Moelker et al. 2005: 31). Hence, there seems a growing mismatch between soldierly requirements and the quantitative or qualitative availability of employees in the security sector.

Since military effectiveness fundamentally depends on the people who serve, the consequences of falling short of the quantitative and/ or qualitative recruitment targets are serious (Dandeker and Mason 2010: 214). For example, the Spanish armed forces experienced very low recruitment and retention rates in the early 2000s despite relaxed screening criteria with regard to educational attainment and cognitive ability. The lack of personnel severely weakened the military deployment capability (Frieyro de Lara 2010: 182f.). The Dutch armed forces experienced a similar manning problem. Constant personnel shortfalls resulted in vacancies and structural under-capacity. As a consequence, Dutch contributions to the international military intervention in Kosovo in the late 1990s were comparatively short-lived. Likewise, the presence of the Dutch troops in Ethiopia and Eritrea during the so-called UNMEE mission in 2000 was limited to a single period of 6 months due to manpower shortages (Moelker et al. 2005: 25).

These are equally likely scenarios for the German military. The initiated internal restructuring of the Bundeswehr increased the pressure to recruit young, talented personnel from an aging and declining labor force. Until recently, about 40 % of professional soldiers were recruited from the pool of basic service conscripts (Federal Ministry of Defense 2006a: 112). Effective by 01 July 2011, the German parliament suspended conscription in favor of an all-volunteer force. In view of the commitment that "the Bundeswehr will continue to be a conscript force in the future" made in the most recent White Paper (Federal Ministry of Defense 2006a: 10), the suspension of compulsory military service caught most observers by surprise. Conscription had been seen as an "indispensable element" of Germany's basic foreign policy orientation and "belonged to the political DNA of the republic" (Schmid 2011). The 2011 Defense Policy Guidelines comment the radically changed situation with the statement: "For decades, conscription guaranteed high force levels, augmentation capabilities, and high-quality recruitment [...]. With the suspension of conscription, an important element of recruitment has ceased to exist. Demographic developments make recruitment [more] difficult for the Bundeswehr" (Federal Ministry of Defense 2011: 17). Hence, the transition to an all-volunteer system exacerbates personnel challenges at a time the Bundeswehr is confronted with new international security threats.

1.2 Germany as a Case Study

Germany serves as a case study for other contemporary industrialized societies. The empirical findings offer new insights into a topic that has been thus far inadequately addressed in international affairs, namely the impact of demographic change on strategic calculations (Freedman 1991: 15). In this connection, Germany is an instructive case given its redefined international role, its neglect of demographic research for several decades, its special civil-military relationship, and the consequent sensitivity towards analyses at the security-demography nexus.

As a result of the clouded history and the "disrepute" of the discipline after World War II, scholarly discussion of demographic issues has long been very limited in Germany (Teitelbaum and Winter 1985: 124f.). In the early 1980s, American and French intellectuals already observed the then emerging fertility decline from a strategic perspective and expressed concern about the potential inadequacy of future military manpower recruitment pools. In Germany, similar analyses were less publicized and treated with great discretion (Teitelbaum and Winter 1985: 110). In secrecy, however, the German Ministry of Defense foresaw future recruitment problems and noted that "the base population of draftable age would be insufficient after about 1992 to provide, under present recruitment practices, for the needs of the armed forces" (McIntosh 1983: 189 cited in Teitelbaum and Winter 1985: 125).

In addition, there has been a long lasting neglect of military sociology in the social sciences in Germany, which hindered the creation of theories and methods and resulted in an ad-hoc, piecemeal nature of research commissioned by the military command (Seifert 1992: 7f.). As in the case of demographic research, this disregard of the discipline could originate from the widespread unease about the national military past. In this connection, Lippert (1992: 6) suggests that social scientists' restraint from military issues may result from the common presumption of an "ideological affinity" of the scholar vis-à-vis the object of research. The deprived state of military-related research in Germany may stem from the latent misconception that research into military sociology would contribute to the feasibility of war. However, the doubts about the validity of this argument raised by Lippert (1992: 6) himself are well taken: If it were true, other strands of research, for example on political extremism, should also be marginalized. Rather, Lippert (1992: 7) perceives a "structural incompatibility" between the modes of operation in the scientific community as opposed to the military organization. Whereas Lippert is primarily concerned with the reception and potential negation of research findings on the part of the military, the detected incompatibility also manifests in the different approaches to dealing with data. While military-related information are mostly classified, social science research rests upon the availability of large-scale datasets and individual-level information. This discrepancy further complicates sound research in the field of military sociology, especially in Germany, where data protection is comparatively strict.

Data on military propensity, recruitment and retention are little publicized in Germany and therefore less explored than in other countries, such as the United States. Research into the micro and macro determinants of the military personnel process is rare since there is little publicly obtainable data and little routine. It is difficult to empirically observe and conceptualize all factors at play, and most social scientists do not even try. In this regard, Haltiner and Kümmel (2009: 75) note that individual-level research on soldiers is complicated by the fact that "access to the object of such research, the soldierly individual, is regulated by the military organization".

The inclusion of demographic change as an additional intervening variable makes the present analysis stand out from other studies about military recruitment, yet equally separates it from the existing theoretical frameworks. The reason is that demographic change has rarely been included in the empirically rooted discussion about the drivers of enlistment decision-making, at least in Germany. While U.S.-centered research gives ample attention to the military propensity of future youth cohorts, there appears to have been no systematic review of the demography-induced recruitment challenges of the Bundeswehr. When personnel planners of the Bundeswehr consider demographic trends, they largely neglect the qualitative impact of demographic change on military recruitment in the future. However, the provision for the qualitative dimension of military recruitment, i.e. the human capital endowments of youth willing to enlist, is essential due to, at least, two reasons: Firstly, a pure quantitative analysis of the effects of demographic change on military recruitment underestimates the scale of the looming recruitment challenge. Secondly, a neglect of youths' qualitative characteristics would also be contradictory to the professionalization of the Bundeswehr and the particular importance of human factors in terms of mission effectiveness.

1.3 German Security Policy in Transition

In an assessment of the strategic security environment, the German Ministry of Defense identified a range of risks and threats to international security and the national interests of Germany. Accordingly, the most prominent drivers of insecurity include globalization, ensuing power shifts, state failure, international terrorism, climate change and other natural disasters, resource scarcity, epidemics and pandemics, and the disruption of critical infrastructure (Federal Ministry of Defense 2011: 1–3). The issue of population and security is recognized with regard to uneven population densities, worldwide migration and potential ramifications for regional stability and border security (Federal Ministry of Defense 2011: 4). However, with this, the governmental view on the strategic environment neglects the security risks of population aging and decline. In contrast, the authors of the 2008 U.S. National Defense Strategy took a more progressive approach and urged to take into account both, the implications of "population growth in much of the developing world and the population deficit in much of the developed world" (U.S. Department of Defense 2008: 5).

In light of the identified security risks, German Defense Minister Thomas de Maizière prepared soldiers and the general public for greater international responsibilities of the Bundeswehr and a higher frequency of deployments. The minister argued that Germany would have to increase its engagement in military operations led by the United Nations, NATO, or the European Union given that, thus far, Germany neither lived up to the expectations of its partners nor to its economic weight. The new defense planning assumptions envisage that the Bundeswehr in the future will be sized and shaped to conduct two major stabilization operations and several minor interventions at the same time. In order to improve operational readiness, the target number of sustainably deployable military personnel was raised from 7,000 to 10,000 soldiers (Maizière 2011).

The core tasks of the Bundeswehr were outlined in the 2006 White Paper and reconfirmed in the 2011 Defense Policy Guidelines. The level of ambition for the Bundeswehr comprises a list of contributions to international conflict prevention and crisis management within the framework of collective defense, including the capabilities to assume command responsibility as a framework nation and provide the required capabilities for the entire task spectrum. The intensity scale covers observer missions, advisory and training support, as well as preventive security measures. In addition, the Bundeswehr has to fulfill tasks of homeland security and territorial defense. The outlined national level of ambition for the Bundeswehr defines the quantity and quality of the capabilities to be provided, in particular the requirements for personnel, equipment and financial resources (Federal Ministry of Defense 2011: 10–14).

The wide range of responsibilities associated with the comprehensive management of risks relevant to German security places a serious strain on the Bundeswehr. Meanwhile, its coping capacity is on the decline due to the downsizing and organizational restructuring after the abolishment of conscription. In the foreseeable future, the emerging scarcity of military personnel and other resources will likely exacerbate due to the demographic changes underway.

1.4   New Constraints of Demographic Change

Germany is an illustrative example of the demographic changes under way in most advanced industrial countries (Leahy et al. 2007: 57f.). Life expectancy is still rising. Fertility rates are below the replacement level. There are significant changes in the population age structure. These demographic trends play out quite differently across regions. The ethnic heterogeneity and cultural diversity are increasing (Hullen 2004; Vaupel and von Kistowski 2008; Höhn et al. 2007).

Demographic trends are an important variable for military planning. They inform about the number of people at recruitable age and, hence, about the sustainability of future military recruitment. Estimates of the future size and age structure of the population are equally important for a preliminary assessment of public preferences and fiscal trends. Both, a public orientation towards inward-looking domestic policies and greater budgetary pressures due to increased age-related spending, may constrain the capacity of aging states like Germany to conduct foreign relations and pursue their national security interests. The following paragraphs provide an overview of the main demographic drivers that are critical for an assessment of the demographic impact on Germany's security policy.

1.4.1   Mortality

Especially since the mid-1950s, life expectancy in Germany has been on the rise, albeit on different levels in the East and West. After reunification, life expectancy in East Germany increased at an accelerated rate, phasing down the East–west gap in life expectancy. Current gains in life expectancy are mainly driven by improved survival at older ages (Gjonça et al. 2000: 2–16). In view of the decline in disability at older ages, there is reason to believe that the gained years will also be spent in better health (Manton et al. 1997: 154; Ziegler and Doblhammer 2005: 16).

According to estimates by the Federal Statistical Office, average life expectancy at birth currently stands at 82.6 years for females and 77.5 for males. Remaining life expectancy at age 65 is 20.6 and 17.3 years respectively (Federal Statistical Office 2011b). The 12th coordinated population projection assumes a continuous increase in life expectancy in the coming decades, however less rapidly than in the recent past. According to the baseline scenario, average life expectancy will reach 89.2 years for females and 85.0 years for males by 2060, with the longevity gap between females and males slowly converging (Federal Statistical Office 2009: 29f.).

These trends in health and old-age survival are indirectly relevant to the foreign policy capacity of states because they provide information on age-related spending, and therefore, the budgetary room for foreign and security policy.

1.4.2 Fertility

Germany's low-fertility regime is similar to other contemporary industrialized countries. Drivers of the decline in fertility to sub-replacement levels include the postponement of marriage and childbearing, the decline in higher-order birth rates, and a higher incidence of voluntary childlessness (van de Kaa 1997: 10).

In both East and West Germany, important factors associated with low fertility emerged around 1965, when birth control was liberalized and societal values changed in favor of self-actualization, and a pluralization of family forms (Pötzsch 2007: 8; van de Kaa 1987: 11f.). Albeit the parallel decline in fertility rates, the process of the fertility transition was quite different in East and West Germany, particularly after the mid-1970s (Höhn et al. 2007: 11ff.). Ever since, West German fertility stabilized at a level around 1.3–1.4 children per woman (Kreyenfeld and Konietzka 2008: 52). However, the share of childless women increased from 21 % among those born in 1955 to 28 % for the cohort born in 1966. By contrast, childlessness in East Germany only reached a level of 10 % among women born in 1965. In addition, a fairly high share of East German couples had two children (Dorbritz and Ruckdeschel 2007: 50). However, trends in East German fertility have been relatively volatile over time. After reunification, the average number of children per woman dropped by more than a half from 1.52 in 1990 to 0.77 in 1994. Thereafter, fertility rates picked up and stabilized at a level similar to that in West Germany. In 2007, there was only a slight variation in the average number of children per woman in East and West Germany (1,366 vs. 1,375 children). Three years later, the fertility rate in East Germany (1,459 children) was slightly, yet noticeably, higher than in West Germany (1,385 children) (Federal Statistical Office 2011a).

Projections of fertility are associated with a high level of uncertainty. In contrast to projections of life expectancy, historical trends are less indicative of the future. In part, this also explains the existing disagreement in the scholarly debate. For example, Dorbritz and Schwarz (1996: 255f.) argue that a significant reorientation towards the family is unrealistic. They expect that individualistic lifestyles (with few or without children) will prevail and that fertility rates will remain low. In contrast, Konietzka and Kreyenfeld (2008: 70) refer to the important role of family policy to alleviate institutional barriers to childbearing, enhance the familial orientation in society and raise the fertility rate. In view of recent fertility increases in Germany, there are some signs that support the latter view.

Fertility trends are relevant to foreign policymaking because they inform about the size of the future military recruitment potential. Changes of reproductive behavior influence the relative cohort size and alter the population age structure.

Table 1.1 Total population and 18 year-old population, 2010–2060

	2010	2020	2030	2040	2050	2060
Total population (million)	81.5	79.9	77.4	73.8	69.4	64.7
Relative change (%)	*100*	*98*	*95*	*91*	*85*	*79*
18-year population (thousands)	848	733	676	667	586	536
Relative change (%)	*100*	*86*	*80*	*79*	*69*	*63*

Source: 12th coordinated population projection (Federal Statistical Office 2009)

1.4.3   Changes in the Population Age Structure

Demographic trends in Germany are governed by a significant upward shift in the population age structure, i.e. a sustained increase of the demographic weight of the elderly and a relative decline of younger age groups, and the subsequent decline in the population size. According to current estimates, Germany's aging process will peak in 2030 when the baby boomer generation, a huge bulge in the population pyramid, will enter retirement age. Thereafter, the dynamic of the aging process will abate, and the main trend will be population decline (Dorbritz et al. 2008: 11).

Based on the 12th coordinated population projection of the Federal Statistical Office, Germany's population will continue to age and shrink over the next few decades (Table 1.1). In reference to the baseline scenario, which rests upon moderate assumptions about fertility and mortality and an annual net immigration of 100,000 people, the total population will decline from 81.5 million people in 2010 to about 64.7 million in 2060 (−21 %). At the same time, the youth population will contract many times over. For example, the size of the 18-year population will shrink from its current size of 848,000 to 536,000 in 2060 (−37 %).

Figure 1.1 displays the shift in the relative demographic weight of younger and older age groups in Germany between 2010 and 2060. It demonstrates the gradual decline in children, youth of recruitable age, and people of working age, and the increase in the population at retirement age.

Depending on the demographic weight of one age cohort, changes in relative cohort size may influence political majorities in public elections. In view of their sheer cohort size, the elderly will have a relatively large voting power. At the same time, birth cohort changes affect the size of several labor market groups and may improve the career prospects and earning abilities of the young population. A smaller cohort size, due to constantly low fertility levels, means a supply reduction in human capital and thereby increases the competition for young and talented employees. This would also make the fulfillment of the quantitative and qualitative recruitment targets of Western militaries, like Germany's, more difficult.

1.4.4   Regional Differences

Aggregate trends conceal some distinct regional patterns in population size and age structure. While East and West Germany are faced with a similar demographic

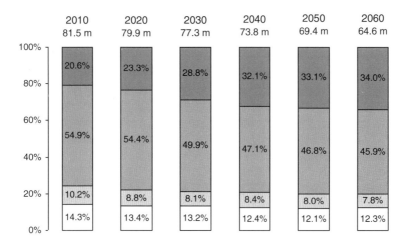

Fig. 1.1 Change in the relative shares of age groups, 2010–2060 (Source: 12th coordinated population projection (Federal Statistical Office 2009))

future, the speed and scope of demographic change will be more dramatic in East Germany (Mai 2008: 291). The divide between a rapidly aging, shrinking population in East Germany and an aging, yet slightly growing population in West Germany is already manifest since the early 1990s. The major reasons for such unequal trends included very low fertility and strong out-migration in East Germany as opposed to substantial net migration gains in West Germany, which compensated for the natural decrease (Höhn et al. 2007: 27f.). The significance of this regional demographic divide is documented by a number of recent reports and projections. For example, a 2008-based regional population projection of the statistical office of the European Union shows that the demographic challenge in Germany will be unmatched anywhere else in Europe, particularly with regard to East Germany. Hence, seven of the ten regions in Europe with the highest median age – now and in the future – are situated in East Germany, including Mecklenburg-West Pomerania, Brandenburg (Southwest, Northeast), Thuringia, Dresden, Saxony-Anhalt and Chemnitz (Giannakouris 2010).

In the future, the demographic differences between East and West Germany will lessen. According to a recent spatial projection, there will emerge a wide corridor of regions in Germany by 2030, with either a stagnating or a shrinking population (Federal Ministry of Transport and Urban Development 2009). This corridor spans from the north of Hesse to the southeast of Lower Saxony and extends to Mecklenburg in the North and to the German-Czech border near the Danube River in the South. In other regions, the population will age less rapidly and still slightly grow. However, the number and scope of these growth regions will decline over time. Population growth will be limited to a few attached areas in West Germany, in particular the Munich area, as well as both Hanseatic cities Hamburg and Bremen. Other regions

Table 1.2 Dynamics of demographic change in Germany, 2006–2050

	2006	2010	2020	2030	2040	2050
East Germany (without Berlin)						
Total population (million)	13.2	12.9	12.0	11.0	10.1	9.1
Relative change (%)	*100*	*98*	*91*	*83*	*77*	*69*
18-year males (thousands)	93.5	40.6	49.3	47.6	38.3	34.5
Relative change (%)	*100*	*43*	*53*	*51*	*41*	*37*
18-year females (thousands)	87.6	36.9	45.9	45.1	36.4	32.8
Relative change (%)	*100*	*42*	*52*	*51*	*42*	*37*
West Germany						
Total population (million)	65.7	65.6	64.7	62.9	60.2	56.7
Relative change (%)	*100*	*100*	*98*	*96*	*92*	*86*
18-year males (thousands)	394.4	381.9	318.2	284.0	281.4	248.8
Relative change (%)	*100*	*97*	*81*	*72*	*71*	*63*
18-year females (thousands)	375.2	366.9	303.9	269.8	267.1	236.2
Relative change (%)	*100*	*98*	*81*	*72*	*71*	*63*

Source: 11th coordinated population projection, regional variant (Federal Statistical Office 2006a)

are projected to experience at least some demographic growth, including individual areas in North-Rhine Westphalia (particularly along the Rhine River) and in Baden-Wuerttemberg (Middle-Neckar Region, southern Black Forrest, Lake of Constance). Those regions with a shrinking population will also be experiencing a significant decline in youth. In East Germany, especially in the peripheral, economically weak regions, this decline will be dramatic.

Table 1.2 illustrates the East–west divide in population dynamics numerically. Three trends are noteworthy: Firstly, the relative decline in total population size progresses at a higher rate and larger scale in East Germany. For example, while by 2030, East Germany's population size is projected to shrink to 83 % of its size in 2006, the West German population will still be at 96 %. By 2050, the East German population is projected to decline to 69 %, while the West German population will remain at 86 % of its initial size. Secondly, in both parts of Germany, the decline of the youth population will be more severe than that of the total population. Thirdly, as a result of the demographic distortions after reunification, the size of the youth population in East Germany currently plummets, and the population group of 18 year-old males fell from 93,500 to 40,600 (equal to 43 % of the initial size) within the short period from 2006 to 2010. By 2030, the number of 18 year-old males is projected to rise again to the level of 47.600 (equal to 51 % of the 2006 size), yet in the longer run, it will be less than half of its size in 2006.[1] In the meantime, the decline in the West German youth population is less dramatic. From

[1] Despite the availability of the 12th coordinated population projection, the estimates of the preceding 11th coordinated population projection are displayed here. This is to illustrate the demographic effect of the sharp fall in the birth rate in East Germany during reunification, i.e. between 1988 and 1992. With a base year of 2010, as in the 12th coordinated population projection, the actual reduction would have been concealed.

2006 to 2030, the 18 year-old male population is projected to fall from 394,400 to 284,000 (equal to 72 % of the initial size).

Divergence in regional demographic patterns is relevant to military personnel planning in view of the variation in military propensity and military enlistment. In Germany, signs are that regions with traditionally high enlistment rates, like in structurally weak, rural peripheral regions in East Germany, will be confronted with a disproportionate decline in the youth population.

1.4.5 Ethnocultural Diversity

The German population has become ethnically and culturally more diverse due to the large-scale recruitment of foreign labor, in particular from Southeast Europe in the 1960s, the uptake of asylum seekers or political refugees from crisis regions, such as the Balkan in the early 1990s, and the influx of ethnic Germans, especially from Eastern Europe since the late 1980s, (Hullen 2004: 18f.; Höhn et al. 2007: 14f.). Principally, this development is observable in all regions of Germany. However, migration streams usually follow existing networks and therefore concentrate on metropolitan areas in the West and South of Germany, as well as Hamburg and Bremen. In East Germany, increases in the relative population share of migrants are only recorded in a few central cities, such as Berlin, Leipzig, Dresden, Halle, and Rostock. These differences in immigration patterns across Germany will also have a lasting effect on the region-specific rates of internationalization in the future. According to the most recent spatial projection, international immigration will be largely restricted to major cities. There, it will coincide with an already slower process of demographic aging (Federal Ministry of Transport and Urban Development 2009).

As a result of these trends, the workforce will become more international. This applies especially to the new entrants into the labor market. In view of the major demographic changes taking place in many of the major sending countries in Eastern and Southern Europe, the origin of potential migrants in the future will be increasingly situated outside of Europe, especially in Africa (Coleman 2006: 415).

In light of this new face of migration, observers expect the religious makeup of Western European countries to change as well. For example, Goujon et al. (2007) project the share of the Muslim population in Austria to increase from about 4 % in the base year 2001 and to a level of 14–18 % by 2050. By way of illustration, the share of the Muslim population in France currently amounts to about 7–8 %, which is a level that Kaufman (2008: 9) rates as a reasonable and realistic magnitude for all other European countries. In official projections, fertility rates of the foreign-background population are assumed to remain higher than that of the resident population for two generations (Coleman 2012: 179ff.). Hence, natural growth in the foreign-origin population and major migration streams from other parts of the world than today will significantly alter the social fabric of contemporary European societies and increase diversity (Kaufmann 2008; Coleman 2012).

In view of the mandatory requirement for soldiers in the Bundeswehr to hold German citizenship, relative changes in the ethnic and cultural makeup of the population are important parameters in the estimation of the future military recruitment potential. A failure to observe differences in fertility rates, and therefore demographic trends, between ethnic groups may lead to an incorrect assessment of the eligible youth population. In Germany, this would mean an overestimation of the actual military recruitment potential since a significant share of the youth population does not possess German citizenship.

1.5   Research Framework and Outline

In view of the continuing military recruitment problems, there is the question about the determinants of adequate military manpower supply and how they will change in the foreseeable future due to demographic change. This research may be of importance to long-range foreign policy decision-making. It proceeds from the observation that Western militaries, like the Bundeswehr, operate in an uncertain environment in terms of future mission requirements and future manpower supply. The underlying assumption is that the social environment of the Bundeswehr should be observed to determine factors that can be directly influenced, anticipate future developments, and adapt policy-making accordingly. The analytical toolbox is interdisciplinary. It draws on theories, data and concepts of demography, sociology, economics and foreign affairs. In that, trends are observed through an "objectivistic" rather than an "interpretative" lens (Carlsnaes 2006: 335).

Following this introduction, the analysis begins, in Chap. 2, with a discussion of the broader military environment, examining economic, technological, sociocultural, and geostrategic developments and how they relate to contemporary military missions and the military organization. Chapter 3 links some of the key demographic trends to emerging security risks and the ability of Western states to provide for security capacity. Chapter 4 takes up the selectivity of military manpower demand and supply in terms of socio-demographic traits and several other characteristics that are valued in the labor market. The empirical analysis considers both basic military service recruits and professional soldiers relative to their civilian counterparts, on the basis of annual survey data from the German Microcensus between 1998 and 2005. Chapter 5 offers an in-depth study of determinants of enlistment propensity and actual military service drawing upon two recent and representative data sources: the 2006 youth opinion poll on the security political situation in Germany and the German Microcensus, an annual, cross-sectional survey of 1 % of German households. Chapter 6 explores the socio-demographic context for recruitment and illustrates that the size of the youth population and the fixation of adverse health effects, including overweight and smoking, will fundamentally constrain military recruitment in the future. Chapter 7 concludes with a summary of the results and a brief discussion of policy options to alleviate the continuing recruitment challenges and secure military readiness in the future.

In view of the societal misgivings of the military past, the predominance of welfare issues in the contemporary analysis of the policy implications of demographic change, the limited research at the security-demography nexus, and ultimately the problematic access of military-related data, the present analysis makes a significant, interdisciplinary and overdue contribution to a less explored field of the social sciences. The ultimate objective is to sensitize the foreign policy community, especially in the military, to consider demography as an important shaper of both security risks and security capacity. Furthermore, the intention is to round out the existing, impartial analyses of the recruitment challenges of the Bundeswehr, and to advance the interdisciplinary dialogue between demographic research and international relations.

Chapter 2
The Military Environment

2.1 The Economic Environment

In the external environment of the military, the defining variable is almost certainly the process of globalization, which refers to the increased global connectedness in political, economic, military and cultural relations. It influences the nature of contemporary conflict and the character of political authority (Kaldor 2007: 4f.).

Economic globalization has created a global marketplace that is supplied by multinational corporations with worldwide production and distribution networks (Held et al. 1999: 236f.). There is a high level of interdependency between national economies, while global trade significantly impacts the welfare and industrial structure of nation states (Held et al. 1999: 178f.). This expansion of global economic interactions aids threats that fall into the "wider security context" (Buzan et al. 1993: 21ff.) of economic, societal and political security. For example, the rising intensity and extensity of global trade and finance has increased the vulnerability of nation states to crises in remote parts of the world. Thus, most states, and the major powers in particular, are forced to be more sensitive to military developments and other security-relevant trends around the world. The character of these sensitivities is usually selective and depends on national interests and the strategic importance of the regions at stake (Held et al. 1999: 102). When the economic security of states is concerned, Buzan et al. (1993) differentiate between the "investment security" (104) and "security of supply" (116).

The economic literature offers two principal notions of "foreign direct investment" (FDI). The first concept centers on international capital flows and the value of asset holdings in companies overseas. The second form of direct investment is that firms shift a set of their economic activities such as production, purchase, sales or other intermediate operations to a foreign host country (Lipsey 2007: 334f.). The attractiveness of states as recipients of capital flows or locations for production derives from the availability and price of local labor, and most of all, their physical, legal and institutional infrastructure. Such enabling conditions include good governance, efficiency in legal order, accessibility of roads and other transportation networks, as reliability in electricity and telecommunications (Carr et al. 2007: 387).

W. Apt, *Germany's New Security Demographics: Military Recruitment in the Era of Population Aging*, Demographic Research Monographs, DOI 10.1007/978-94-007-6964-9_2, © Springer Science+Business Media Dordrecht 2014

FDI inflows into developing countries have been increasing steadily and reaching record levels in the past few years. Regions formerly marginalized in foreign investment, such as sub-Saharan Africa, South and South-East Asia, recorded significant increases, just like the largest recipient countries in Latin America, namely Brazil, Chile, Colombia, and Peru (UNCTAD 2008: 8). Among developed countries, investment outflows exceed inflows. The largest sources of foreign investment are currently the United States, the United Kingdom, France, Germany, Spain, Italy and Japan with outflows of more than $100 billion each (UNCTAD 2008: 75). As a result, taking Germany as an example, there is a significant international diversification of economic activities. When measured by geographical spread or the number of host countries for foreign affiliates, 5 of the 15 largest transnational corporations are German. The number of host countries of such companies like Deutsche Post, Siemens, BASF, Linde and Bayer ranges between 71 and 111 (UNCTAD 2008: 28).

Demographic aging will entail two major consequences for global capital flows. Firstly, it will lead to an increasing relocation of production facilities from developed to developing countries in order to compensate for labor shortages at home and reap lower labor costs in host countries (Nyce and Schieber 2005: 285). Given that China, as the currently main recipient of FDI is aging rapidly, it is plausible that FDI will be increasingly directed to populous countries states "at risk" such as Nigeria and Bangladesh that have thus far entertained weak links to the major economies. However even if countries are politically failed or otherwise unattractive for foreign investment, they are still relevant for the state of the global economy by influencing the accessibility of trade routes, the economic climate for investments in the region, and as suppliers of scarce natural resources such as fossil fuels or timber (Demeny and McNicoll 2006: 277). Secondly, for the time after 2025, Brooks (2000: 29) projects a growing capital inflow from developing countries into industrialized nations, which will likely experience a serious decline in their savings as a result of demographic aging and the foreseeable large-scale disinvestment to finance consumption.

However, globalization is not restricted to the civilian economy. There is also the phenomenon of "military globalization", which has been conceptualized as the intensification of military connectedness among the political units of the world system (Held et al. 1999: 88). The main characteristics of this development include, firstly, the global reach of the war system in terms of the spread and incidence of military interventions, and secondly, the evolution of a so-called "geo-governance" of security affairs that is reflected in the increasing number of international alliance memberships and military cooperation agreements, which regulate the structuring, purpose and use of military power (Held et al. 1999: 89).

2.2   The Geostrategic Environment

With the end of the Cold War, hopes were high for a "new world order", which, in 1990, former U.S. president George H. W. Bush envisioned as "a new era – freer from the threat of terror, stronger in the pursuit of justice, and more secure in the quest for peace, an era in which the nations of the worlds, East and West, North and South, can prosper and live in harmony" (cited in Brown 2003: 2). Several of the world's major powers, in particular the United States, signaled their willingness to cut back on defense expenditures and reduce force levels. This raised expectations for a "peace dividend", i.e. a reduction in military spending over time with potential benefits for the economy and the public good of all states (Mintz and Huang 1990: 1283). However, while the security of the developed world largely benefited, the developing world continued to experience major armed conflicts.[1] The anticipated welfare gains from improvements in international security did not materialize (Chan 1995: 57). One major reason lay in an altered threat perception among Western security elites that brought previously neglected security challenges on the agenda.[2]

The new focus on intrastate conflicts at remote locations evolved from the lower propensity of conventional interstate wars, a broader definition of security, an increasing awareness of the human costs of state failure, and the assumption of an ethical responsibility to prevent or at least curtail such conflicts (Hoyt 2003: 217). As an additional stimulus, the accelerated globalization of politics, economics, culture, and moral principles prompted the international community to engage in the resolution of regional conflicts. In this connection, McGuire (2000: 21) emphasizes two rationales: Firstly, given the industrialized states' dependency on raw materials and global trade, crisis intervention may be justified on grounds of supply shortages of critical resources or the potentially detrimental consequences of conflict-induced trade barriers for the domestic economy. Secondly, Western publics have shown a distinct sensitivity to the atrocities of war and other humanitarian crises, which provided a major impetus for intervention and broadened the spectrum of military tasks. The fear of spillover effects or repercussions for international security constituted another rationale for heightened attention to humanitarian crises, refugee flows, and internal conflicts or political violence in other parts of the world (Edmunds 2006: 1067; Hoyt 2003: 213). Some precedent-setting crises that provoked international military responses during the 1990s included Afghanistan, Lebanon, Somalia, Zaire, and in particular Yugoslavia (Hoyt 2003: 217). The nature of these conflicts was unknown: protracted and

---

[1] In the period after the Second World War (1946–2003), a total of 229 armed conflicts in 148 locations around the world had been documented. Of these, 116 conflicts in 78 areas remained active until after the end of the Cold War (1989–2003). Over the past decade, the total number of major armed conflicts declined from 21 in 1999 to 16 in 2008. The conflicts were located in Africa (Burundi, Somalia, Sudan), the Americas (Colombia, Peru, USA), Asia (Afghanistan, India/Kashmir, Myanmar, Pakistan, Philippines, Philippines, Sri Lanka) the Middle East (Iraq, Israel/Palestine, Turkey/Kurdistan). However, the downward trend in conflict propensity has not been uniform over time, with major drops in 2002 and 2004, compared to increases in 2005 and 2008 (Harbom and Wallensteen 2009).

[2] Some of those chiefly internal wars that broke out before the end of the Cold War include the armed resistance against the declaration of independence in Myanmar/Burma (1948), turmoil associated with the division of India (1948), the Eritrean war (1961), the civil war in Sudan (early 1970s), as well as some other regional conflicts in the Middle East involving Israel, Egypt, Jordan, Syria and Iraq (Harbom and Wallensteen 2009).

low-intensity wars of attrition that lacked the sophistication and the well-defined opponent of earlier conflicts (Hoyt 2003: 217).

Contemporary wars, although sharing many characteristics with earlier low intensity conflicts, have numerous distinctive features that justify the "new war" argument. These features include a strict avoidance of head-on collisions, a great emphasis on surprise and mobility, a decentralized and uncontainable character of violence, an obscure number of warring factions, of whom most are non-state actors, the pursuit of ideological and political objectives, a concentration of conflicts in Africa, Eastern Europe and Asia, and a multilateral, deeply politicized international response based on international norms and humanitarian values (Kaldor 2007: 3–10).[3] The underlying motive of contemporary low intensity conflicts is to target the social-political system of a society, which is in sharp contrast to the conventional and Clausewitzian notion that "the center of gravity in war is the defeat and destruction of the enemy armed forces" (Sarkesian 1985: 5). Hence, the objective of revolutionary forces is to bring territory under social-political control by employing "political, psychological, and economic techniques of intimidation" (Kaldor 2007: 9) and homogenizing people living in the controlled territory by means of expulsion and population displacement or what Weiner and Teitelbaum (2001: 66) call "population unmixing", as seen in Rwanda, Bosnia-Herzegovina, and the Transcaucasus region.

Along with the fading distinction between categories of violence and combatants or non-combatants, recent conflicts showed a fast-rising toll of civilian casualties. The hostilities in Afghanistan, Darfur (Sudan), Iraq and Pakistan have been among the deadliest and fiercest armed conflicts of the recent past (Stepanova 2008). The smoldering conflict in the Darfur region may be considered as a potent example of the complex interplay of demographic, religious, and resource factors in combination with state weakness leading to armed hostilities (summarized in Stepanova 2008). In 2008, the dispute was characterized by a shift from state-based fighting to a compound mix of less intensive but numerous "mini conflicts". The conflicting parties swiftly changed alliances, looted natural resources and public goods, assumed political control of the population and launched transborder attacks to ignite fear and further destabilize the region. The human toll of tribal and factional violence was greater than the number of deaths in the battle between rebels and the government.

Military intervention in contemporary conflict is mandated by multilateral alliance regimes, such as NATO, the European Union or the United Nations. They regulate national military autonomy and constitute a formal agreement between member states about the purpose and deployment of military force (Russett 1971: 262–263). Such a collectivization of security affairs involves opportunities and risks. While it allows for burden-sharing and the downsizing of national militaries, complications emerge if union members disagree about the implementation of

---

[3] Similarly useful as the account on "old wars" and "new wars" by Kaldor (2007: 1–32) is the detailed description of the changing pattern of violence before and after the end of the Cold War offered by Snow (1999: 96–126), in which he focuses on the major characteristics of the "new internal wars".

collective security rules, the appropriateness of collective deterrence and the use of force (Frederking 2003: 372). Hence, under the umbrella of an institutionalized alliance arrangement, a member state may be requested to provide military capacity on behalf of a collective security objective although the intervention may not directly serve the strategic interests of the individual state and/or the domestic public may fail to grasp its significance for national security. In light of varying security objectives and the absence of a coercive authority, the national contributions to security alliances have been very unequal.

2.3 The Technological Environment

Another major factor in the military environment is technology. The enormous developments in military technology, especially since the end of the Cold War, have been coined by a "revolution in military affairs" (cf. Krepinevich 1994; Cohen 1996). Scholars disagree in their understanding of this revolutionary change in warfare. Krepinevich (1994: 32) focuses on the technical-engineering innovations. Hundley (1999: 23f.) emphasizes the evolutionary aspect and argues that the so-called revolution is, in fact, the result of "multiple innovations" in the realm of technology, operational capability, operational concept, military doctrine and force structure. Kaldor (2007: 4) refers to the consequences for the conduct of war and maintains that "it is a revolution in the social relations of warfare, not in technology, even though the changes in social relations are influenced by and make use of technology." This modified interpretation appears like a critical response to the long held view of technological superiority as a "panacea for pressing military problems" (Handel 1981: 227).

Then again, technological progress is driven by social and political factors. Socio-political pressures also shaped the development of military technology. For example, public expectations regarding low combat causalities on either side of the conflict lead to advances in the accuracy of basic weaponry, in particular air power (Hoyt 2003: 27). Similarly, the presence of international media directly at the battleground has altered the conduct of war and pressurized the armed forces regarding which technology should be used on a mission and the professional behavior of the armed forces (Smith 2005: 497).

While the increasing sophistication of weapon technology seeks to improve military effectiveness, it may actually entail the converse paradox of "structural disarmament" (Boëne 2003: 170), which refers to the reluctance of military managers to employ high value equipment when they perceive the gain at stake to be much smaller than the risk of losing scarce weapon systems. For example, Luttwak (2003: 113) cited an incident during the last Balkan War, where, in March 1999, NATO commanders refused to allocate 24 Apache helicopters to protect the persecuted Kosovo-Albanian village population against small troops of Serbian insurgents. While the risk to lose significant material equipment appeared to have played a major role in the decision, casualty aversion likely constituted an additional major concern.

Another implication of the "structural disarmament" paradox are the soaring investment costs for new military technology. With each new generation of weapons, Western militaries purchased lower volumes of procurement items. To some extent, the fact that weaponry is not replaced one-for-one is due to increased equipment efficiency, yet more importantly, it results from "peace dividend pressures" and the associated reductions in post-Cold War defense spending (Boëne 2003: 170). With lower production volumes, it will become much more difficult for the defense industry to achieve economies of scale. As an economic corollary, the prices of defense equipment are therefore likely to increase further. Hence, Alexander and Garden (2001: 509) infer that adequately equipping the armed force is becoming increasingly difficult for the majority of industrialized states. They center their argument on the "arithmetic of defense policy" and refer to the fast rate of increase in personnel costs, equipment costs, and operating costs that will most likely lead to a decline in military capacity even if defense spending levels remain constant (Alexander and Garden 2001: 517). As a countermeasure, there has been a trend towards equipping the military forces with civilian technology. The accelerated speed of technological innovation, the shorter duration of product life cycles in the civilian economy, and the hardening constraints in the defense budget support this development (Manigart 2003: 324; van Crefeld 2008: 246).

In the long run, the outlined aspects of technological change raise major questions about the future modes of warfare and the future shape of the international system. Recent trends suggest that two main technology-relevant security risks will become manifest: Firstly, the rapid diffusion of generic technological innovation will likely adjust imbalances in the battlefield, especially in the context of unconventional warfare and asymmetric conflicts (Hoyt 2003: 27ff.). Secondly, the continued price increase in defense technology and parallel budgetary pressures will make it more difficult for states to maintain the general purpose of their armed forces and acquire high level capabilities in all technological and military niches (Hoyt 2003: 31). Major countries like France, Germany and the United Kingdom already have difficulty to provide their armed forces with the full range of defense equipment. In sum, the cost of conventional equipment and the available financial scope for defense expenditures will divide the international system into different levels of power (Waltz 1979: 183). Similarly, Hoyt (2003: 31) predicts that "[t]here will be wide disparities in military capability based on different concepts and technologies".

2.4 The Socio-Cultural Environment

Domestic political variables have long been neglected in the leading theories of international conflict (Levy 1988: 653). In the meantime, there has been greater recognition of the role of domestic factors in the foreign policy orientation of states.

In the sense, that politicians base their foreign policy decisions on both international *and* domestic concerns, international affairs are understood as "linkage politics" (Miller 1995: 764). Domestic forces may either enable or constrain governmental capacity to carry out foreign policy and, if necessary, project military power (Clark and Hart 2003: 58). In line with this, Dalton (1988 cited in Kernic et al. 2002: 39) argues that public opinion limits the range of acceptable policy options, within which politicians must find ways to resolve pressing issues and conciliate conflicting interests. This system of beliefs, i.e. the national strategic culture, shapes the world outlook of the citizenry and creates a collective bounded rationality that "limits attention to less than the full spectrum of logically possible behaviors, affects the appearance and interpretation of these possible behaviors, and influences the formulation and evaluation of different policy options" (Dalgaard-Nielsen 2006: 10). Following this, Moravcsik (1997: 518) argues that state preferences are most important in world affairs and, in fact, outweigh capabilities or institutional arrangements in significance: "Societal ideas, interests, and institutions influence state behavior by shaping state preferences, that is, the fundamental social purposes underlying the strategic calculations of governments".

Consistent with the argument that domestic politics are a key factor in decisions to use force, the legitimization of military interventions, either for war-fighting or peacekeeping purposes, has proved publicly and politically contentious in most European states. As a result, some countries, such as Austria, Finland, Ireland, Russia and Sweden, oppose any intervention without approval by the UN Security Council and practically limit their military role to peacekeeping and humanitarian interventions. Others, including Denmark, France, Poland and the United Kingdom, have shown readiness to override the territorial integrity of nation states to provide military aid from outside as circumstances demand. However, even those states, that are typically more sympathetic towards the use of military force for war-fighting purposes, have struggled to maintain and sustain societal support for offensive military operations such as that in Iraq. A third group of countries have relied on UN authority to determine when and under which circumstances, military intervention should take place. At the same time, the governments of these countries, including Belgium, Germany, Greece and Italy, have been very hesitant to deploy military force in any other capacity than humanitarian intervention or post-conflict reconstruction (Forster 2006: 203–206; Edmunds 2006: 1067f.).

While the nature of security threats informs about the extent to which they can be alleviated by human intervention, the subjective perception of societies about the relative prominence of these threats, their probability of occurrence and their potential to cause harm shapes the types of instruments that states may deploy to manage the sources of insecurity (Aldis and Herd 2004: 183). This domestic element in foreign policy decision-making does not necessarily bear upon the objective threat situation and may change over time (Smith 2005: 499). Rather, historical experience, national identity and collective values about international affairs work as a lens on the assessment of the nature and sources of threats (Dalgaard-Nielsen 2006: 9).

Opinion polls show that, since the end of the Cold War, threat perceptions in Europe have significantly shifted from external military threats to broader civilian or social risks (Manigart 2003: 328). At the same time, cross-national differences in foreign policy attitudes and priorities remain. In a comparison of public opinion, Jonas (2008: 163) ascertained a strong preference of the general public in Germany to pursue ideational-normative objectives in foreign policy. On the other end of the spectrum, American public opinion clearly favors material-realist priorities. With their endorsement of both ideational-normative and material-realist motives in the formation of foreign policy, the publics in France and the United Kingdom range between the two extremes. Jonas (2008: 164) infers that ideational-normative objectives as in Germany entail a civilian approach or low-intensity operations to ameliorate crisis situations, while material-realist priorities as in the United States seem to justify military deployment and high-intensity operations.

In an analysis of the scholarly debate of Germany's post-Second World War security, Dalgard-Nielsen (2006: 10) reports "a pervasive culture of anti-militarism in German society" that firstly lead to the evolution of a new self-concept as a civilian power, which relies heavily on political, cultural and economic avenues of international influence rather than military means, and secondly, explains the sustained reluctance to have a greater share in the management of international security albeit increasing external pressures. Similarly, in view of the German participation in the Kosovo and later in Afghanistan, Schmid (2011) concludes that "from the German point of view, military interventions (a) have to be the absolute exception, (b) are only permissible in the most extreme cases, and (c) are exclusively justifiable in moral terms".

The "avoidance of foreign policy per se" (Schmid 2011) in the political discourse coincides with a public concept of security that, according to Fleckenstein (2000: 84) has only few military connotations: "Whether or not someone feels secure has little to do with defense and the Bundeswehr". Traditional military threats, such as the possibility of an attack on Germany or armed conflicts in other world regions, have been repeatedly ranked as the least feared threats to personal safety (Fiebig 2008: 22f.). Instead, the general public is more concerned about socio-economic risks (e.g. unemployment, cuts in social welfare, economic crisis, demographic aging), ecological risks (e.g. environmental destruction, global warming, natural disasters) or internal security risks (e.g. political extremism, increased immigration, terrorist attacks).

This prioritization of non-military security challenges is attended by a public loss of confidence in the ability of the military to counter non-traditional threats to national security (Kernic et al. 2002: 18). According to Shaw (1991: 64), there has been a shift towards a "post-military society" that manifests itself in a growing demilitarization of everyday life. The lower exposure of all segments of society to the military organization and defense sector aids "a relative social devaluation" (Dandeker 1994: 129) of the military profession and its public perception as "just another job".

The shift in the functional role set towards policing and other subsidiary tasks of regulation or administration, mostly at far-off locations, altered the military's social role and status in society. Due to the lessened significance of unrivaled military functions and the diminished integrative role of the military in society, the armed forces have been regarded as "simply being part of the public services like any other state institution" and have been liable to "an instrumental evaluation" on the basis of cost-benefit analyses and efficiency considerations (Haltiner and Klein 2005: 19). In equal measure, Feaver et al. (2005: 240) observe a growing military-related "experience gap" among Western political elites that emerges from the overall decline of the mass army and the socio-economic selectivity of military service. This gap becomes manifest in a looming clash of interests between political and military elites. The shift in societal norms also shows in the diminished tolerance of protracted combat operations in face of the involved risk of human losses, both in military personnel and the civilian population (Rosenau 1994: 40).

Already Huntington (1975 cited in Domke et al. 1983: 19) observed a sustained "welfare shift" in the United States, that involved a shift of mass opinion and public priorities toward securing social goals, which would make it more difficult for national leaders to warrant expenditures in favor of foreign and defense issues. This transformation may be attributed to, firstly, a low salience of security affairs in public life (Risse-Kappen 1991: 481), and, secondly, a societal value shift from "materialist" to "postmaterialist" priorities, which involved a shift from attaching greatest value to economic and physical security, and in doing so, endorsing the pursuit of "traditional symbols of sovereignty and national power" (Domke et al. 1983: 19), towards a stronger endorsement of self-fulfillment, individual freedom and the quality of life (e.g. Inglehart 1997: 131).

The societal emphasis on civil liberties and individual rights contradicts the nature of military service, which removes or, at least, curtails those rights and thereby renders the soldierly profession less attractive to civilians (Smith 2005: 502). In view of the general refusal of heteronomy, obedience and subordination to institutional authorities (Burk 1994: 10), there has been an "authority crisis" that originates from the societal skill revolution and the enlargement of analytic capacities among average individuals. With a general increase in material prosperity and an expansion of educational opportunities, there is a greater range of employment opportunities and lifestyle choices for youth to choose from. At the same time, youth tend to have a better command of modern technology compared to previous generations and therefore observe international affairs more thoroughly. They have commonly adopted a rather critical world view that entails a frequent questioning of traditional authority, a growing incentive to honor subnational values and a greater sense of affiliation with interest-driven movements than with state institutions (Rosenau 1994: 30f.). The related decline in religious belief may work as an additional deterrent to military service since youth may develop an increasing unwillingness to risk their earthly existence (Smith 2005: 503).

In this connection, former Federal President Köhler (2005: 6) asserted that a genuine interest in military affairs or pride of the Bundeswehr are very rare. Even though public acceptance has been high for many years and the vast majority of the population trusts in the Bundeswehr (Bulmahn 2008: 18), the ambition to under-stand, observe or even participate in the military is very low. From this, Köhler

(2005: 6) concluded the widely cited notion of a "friendly disinterest" towards the Bundeswehr. Hence, the high degree of military legitimacy cannot be taken as an indicator of youth's willingness to make a private contribution to public security by enlisting as a soldier. Against this background, Fleckenstein (2000: 88) detects a general attitude of "Yes, but without me!"

The clash between individual attitudes, aspirations and the soldierly job profile may lead to frustration and detract from future military recruitment when disgruntled soldiers report to members of the core target group. The potential impact of such negative word-of-mouth was illustrated by a survey of 45,000 German soldiers. It revealed a daunting discontent with the level of political and societal support, the amount of defense funding and the state of military equipment. Against this background, only 18 % of the surveyed soldiers stated that the Bundeswehr would be able to recruit a sufficient number of quality personnel in the future, whereas 76 % were convinced that there would be recruitment shortfalls. In the end, this notion may become a self-fulfilling prophecy when one considers soldiers' roles as "opinion leaders" or "insiders". Only 34 % of the soldiers in the survey would advise their next of kin (e.g. children) to enlist in the Bundeswehr, while 59 % said they would advise against choosing the Bundeswehr as a profession (Strohmeier 2007: 11).

2.5   Military Missions and Roles

Among Western scholars and politicians, there is a common understanding that the armed forces represent the "key actors in addressing national, global and regional insecurities" (Edmunds 2006: 1061) making them "far from obsolete" in the foreseeable future (Burk 1994: 7). The salience of the military element in contemporary foreign policy is predicated on the understanding that many contemporary security challenges, such as terrorism and trafficking in humans or drugs, have their roots in crisis-ridden areas that are marked by political instability and longstanding turmoil. The restoration of peace and the develop- ment of legitimate and effective security authorities represent one of the key counteractive strategies. The absence of violence and insecurity constitutes the prerequisite for any major progress in political and economic development (Edmunds 2006: 1064).

The altered nature of insecurities necessitated a fundamental reassessment of the functional role set of the military. It was adapted to such a degree as to be only loosely associated with the traditional core function of territorial defense (Edmunds 2006: 1060ff.). However, this did not effect a task substitution. Rather, Western armed forces were charged with an "extended role set", in which newly emphasized demands became "complementary tasks" to the traditional roles aimed at the security and integrity of national territory (von Bredow and Kümmel 1999: 16–18; von Bredow 2007: 176). These functions include expeditionary missions for war-fighting, peace enforcement or peacekeeping, internal security and policing missions, and an increased emphasis on nation-building and domestic military assistance (Edmunds 2006: 1064).[4]

Mackinlay (1994: 159) categorizes the military task profile according to the level of intensity and thereby provides a fitting typology of the contemporary operational spectrum. He differentiates between (1) low-level missions aimed at assertive military presence in volatile security environments, e.g. observer missions under the UN umbrella; (2) mid-level peacekeeping missions that employ a selective set of military means to contain the opposing force and stabilize a post-conflict zone in terms of civil or humanitarian aspects, e.g. the NATO-led Stabilization Force in Bosnia and Herzegovina; and (3) high-level missions that make use of the entire range of military capability to restore a legitimate monopoly of violence and provide the basis for the rehabilitation of social order, e.g. robust interventions in Afghanistan, the Balkan, and Iraq.

According to Dandeker (2003: 408), major interstate conflict and the maximum violence to win a war in the Clausewitzian sense are the least likely mission scenario for Western militaries in the foreseeable future. In contrast, Gray (2007: 245) argues that the past few decades of irregular and low-intensity warfare are not predictive of the future face of conflict: "Thucydidean motives of fear, honor and interest" (281), i.e. realist power politics that entail regular warfare among states, would continue to pose a viable security threat in the future.[5] Nonetheless, military strategist van Creveld (1991: 22) conceives low intensity conflicts that employ asymmetrical warfare and take place in developing countries are "by far the most important form of armed conflict in our time".

According to former British general Smith (2006: 267–305), military operations in such "war[s] amongst the people" entail changing ends like the need to establish conditions for sustainable socio-political development and to win the "hearts and minds" of the local population; an increasing shift of the battlefield into civilian and mostly urban settings; a timelessness or infinity of operations, which results from the modified purpose of war and the strategic objectives of an inferior opponent; a mode of warfare aimed at preserving the force; and the management of conflicts within some form of multinational grouping to share the financial burden, spread the risk of failure and increase legitimacy.

---

[4] Kaldor (2007: 132) outlines the differing military roles in the primary missions of the military to enforce or preserve peace. According to this, peace enforcement has to be mandated by the UN Security Council and basically means fighting in a war on one side. On the other hand, peacekeeping efforts come after a ceasefire and peace agreement between the conflicting factions; the main military task constitutes in the supervision and monitoring of the agreement.

[5] Thus, Gray (2007) follows the traditional foundations of security studies that maintain that states are in a constant security dilemma. As an adherent of a realist interpretation of international relations, he assumes that states need their armed forces in order to be able to use or threat force where their interests are at risk. The military is seen as the necessary "countervailing force" in the relations between states and societies. Realists like the Prussian military thinker Carl von Clausewitz and the Greek philosopher Thucydides state that military dominance would be the key element establishing a balance of power and a situation of sustainable security. In classical realism, the ability to project power is interpreted as the main factor driving state behavior. In this vein, military might is considered as the principal form of power.

Similarly, based on his analysis of the Second Gulf War, Däniker (1992: 165–188) advances seven theses about the nature and role of the armed forces in the twenty-first century. He maintains that: (1) The core military functions are inherently preventive, interventionist and regulatory, while combat and deterrence are of secondary importance; (2) military victory remains a tactical objective, while the true strategic purpose is to restore peace and stability in order to build a foundation for sustainable development inclusive of the former opponent; (3) the end goal of military operations does not consist in the devastation of the enemy, rather in disarmament and reconciliation; (4) military means have to be commensurable and aim at minimizing the number of civilian and war casualties; (5) military thinking increasingly needs to cut across the boundaries of the traditional state-centered, mostly realist world view and threat perception; (6) flexibility and multi-functionality of the force are just as important as combat strength and maneuverability on the ground; and (7) the main soldierly mission is to "protect, assist, and rescue" and foots on a general orientation towards making a contribution to the preservation and restoration of peace, and thereby improving the life circumstances of people in areas of conflict.

In this way, the underlying rationale of contemporary military operations abroad rests upon a wider framework of international security (Buzan et al. 1993; Buzan 1997) and the assumption that the security of Western states may be affected by instability in different regions of the world although there may be no direct physical or territorial threat. However, the contemporary task profile of the military is not necessarily the result of a new international security situation but mainly arises from an altered socio-political perception of how modern forces should be set up and which tasks they should accomplish (Edmunds 2006: 1059).

2.6   Military Organization

The described trends in the broader security environment and the modified functional role set promoted a large-scale transformation among European military organizations (see Fig. 2.1). Due to the unpreparedness of large, heavy armed forces to meet the challenge of protracted low-intensity operations, there has been a trend towards military professionalization. Force structures have become smaller in size, leaner in personnel, increasingly flexible and multi-purpose, technically more advanced, structurally more differentiated, and hence, mostly all-volunteer since the late 1980s (Burk 1994: 10; Edmunds 2006: 1059).

To fulfill their complex task profile, Western militaries have come to rely on well-trained, immediately deployable professional soldiers, and less on compulsory military personnel. The volunteer force format and the decline of the mass army have been enforced by highly selective conscription or the abolishment of compulsory military service altogether (Haltiner 1998: 32f.). The ongoing force transformation and the varying degrees of military professionalization allow for distinguishing four major types of force structures and suggest type-specific changes

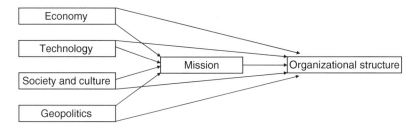

Fig. 2.1 Main determinants of the military organizational structure (Source: Manigart 2003: 324)

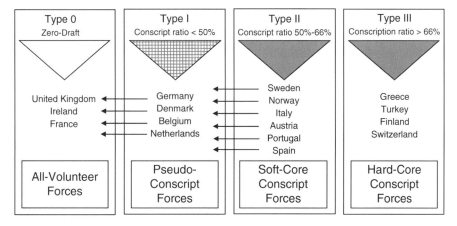

Fig. 2.2 Typology of current force structures and expected changes (Source: Haltiner 1998: 18. In Germany, conscription was formally suspended in July 2011)

(see Fig. 2.2). Until the suspension of conscription in 2011, the Bundeswehr belonged to the group of "pseudo-conscript forces" that bear a strong organizational resemblance to all-volunteer forces and merely differ from traditional zero-draft force types in a draft, mainly used as a "personnel reservoir" (Haltiner 1998: 19).

Taking the early-stage U.S. campaign in Afghanistan as a case study, Biddle (2002) makes some insightful predictions about the future of warfare and appropriate force structures. In reference to the high risks associated with an overly reliance on special operations forces, aerial warfare and precision weapons, Biddle (2002: 50) cautions against the large-scale reduction of ground forces. He argues that "a balanced, all-arms force structure [...] offers important leverage in a world where we cannot know exactly where or with whom we may be forced to fight." Furthermore, he elaborates that an understanding of warfare "as a problem of interactions among armored vehicles and major weapon systems" may jeopardize mission effectiveness by marginalizing soldiers, the all-dominant human element in combat.

Ideal type/ model	Spectrum of conflict	Principal missions	Force doctrine	Defense reform philosophy	Personnel issues	Countries
Expeditionary Warfare	Full range from high to low intensity operations	Warfighting and force projection	Joint and combined	Transformation with no end goal	• Complex command and control networks, flat hierarchical structure • Volunteer recruitment with high qualification levels for entry • Soldierly requirements include physical abilities, technical and managerial skills, constant re-skilling • Recruitment shortfalls due to entry requirements, occupational risk, long and frequent deployments • Low retention particularly in view of occupational stress, slow promotion and competition with civilian sector • Increasing recruitment of women and ethnic minorities	France, United Kingdom
Territorial Defense	Ability to cover tasks in the medium to low spectrum of conflict	Defense of national territory with some limited international security missions, e.g. peacekeeping	All arms	Modernization	• Strong commitment to conscription of about 12 months • Varying military proficiency dependent on funding • Traditional military education system • Achievement of quantitative/ qualitative recruitment/ retention objectives subject to contracts and conditions • Entry level standards declining in Russia and Ukraine, otherwise high • Appraisal of international experience and foreign language skills	Albania, Belarus, Bosnia-Herzegovina, Bulgaria, Croatia, Czech Republic, Estonia, Finland, Georgia, Greece, Hungary, Latvia, Lithuania, Macedonia, Moldova, Norway, Poland, Russia, Serbia-Montenegro, Romania, Slovakia, Slovenia, Sweden, Turkey, Ukraine
Late Modern	Limited aspiration and capability to cover the full range	International security missions, principally peacekeeping	All arms with some limited joint and combined units	Transformation for some units to operate alongside expeditionary forces	• Vertical lines of authority based on command and control • Volunteer recruitment focused on educational attainment • Conscription in Denmark and Germany • Sophisticated military education system similar to expeditionary forces • Low retention among those with specialized skills and field experience • Promotion based on merit, technical skills, performance reviews, international operational experience, foreign language competence	Belgium, Denmark, Germany, Italy, Netherlands, Portugal, Spain
Post-Neutral	Low intensity operations	Defense of national territory with some very limited international security missions, limited to peacekeeping	Militia-based forces	Modernization	• Small officer cadre supplemented by conscripts and national reserve • Inflexible command and control networks, rigid and low-tech equipment • Recruitment aimed at high education, management and budgeting skills • Extensive higher education system, also regarded as incentive • Personnel with limited skill set given low technology and task spectrum • Slow promotion and flat hierarchy given small force size • Retention problems due to low unemployment (e.g. Switzerland)	Austria, Ireland, Switzerland

Fig. 2.3 Key features of ideal type European armed forces (Source: Forster 2006: 67–71)

Although all Western militaries are faced with similar challenges in their external environment, there appears to be no single force model to which European governments subscribe. Rather, as demonstrated by Forster (2006: 67–71), there are major force structure types that emerge in accordance with different perceptions of threats, the varied functional role set of European militaries, the amount of funding and civil-military relations. For the latter, a useful indicator is the ease or difficulty to recruit and retain sufficient quality personnel (see Fig. 2.3).

2.7 Conclusion

An array of trends in the economic, geostrategic, technological and socio-cultural environment of the military influences the role set and organization of the military. There appears a clear trend towards an expansion of the individual and organizational task profile, which places a premium on skilled military personnel and makes recruitment both more important and more challenging. While demographic change unfolds as the main obstacle to military recruitment in the foreseeable future, Western militaries already struggle with filling their ranks today. The main reason for current recruitment shortfalls consists in the mismatch between the applicant profile and job requirements, both in terms of applicants' skills and lifestyle preferences. As the youth population further declines, the quantitative and qualitative recruitment challenge will intensify. Militaries will run short of critical personnel at a time when international security challenges are changing and new sources of conflict appear. In the demographically young and growing countries, large youth bulges approaching the labor market will exacerbate existing scarcities and vulnerabilities. In the assessment of options for Western military intervention, it can be assumed that demographic change and its political, fiscal and social implications will play a greater role in the foreseeable future.

Chapter 3
A Demographic View on Security

3.1 The Global Demographic Divide

In the next decades, the world population will continue to grow. According to the latest United Nations estimates, the size of the world population of about 6.9 billion in 2010 is projected to reach 8.3 billion by 2030 and 9.3 billion by 2050. This growth will, just as in the twentieth century, mainly stem from the natural increase in developing countries, whose population count is expected to increase from 5.6 billion in 2010 to almost 8 billion by 2050 (UN 2011). A significant portion of this increase will take place in the 50 least developed countries of the world whose volume will likely more than double from 832 million in 2010 to 1.7 billion in 2050. As a result, the ten most populous countries in 2050 will be situated in the developing world. With Bangladesh, the Democratic Republic of Congo and Ethiopia, three of the least developed nations will be among them. The populations of the industrialized world will stagnate, roughly stay at its current size of 1.25 billion and grow significantly older (UN 2011).

However, population aging will not remain a phenomenon of the industrialized world. Many of the less developed countries will be confronted with an accelerated aging process. The close connection between economic and demographic development will relax, and the dichotomy of more and less developed nations will be less indicative of the prevailing age structure and future demographic trends (Wilson 2001: 167; Kent and Haub 2005: 3). Hence, the two governing phenomena of the future development of the world population will be the diversification of demographic patterns and global aging.

One major reason behind this development lies in the gradual convergence of mortality and fertility trends. Due to economic progress, social modernization and the ensuing improvements in nutrition, hygiene, and public health, world life expectancy more than doubled since the beginning of the nineteenth century (McMichael et al. 2004; Riley 2001). Still there is some variation in the progress made against mortality. Caselli et al. (2002) classify three groups that display similar dynamics: (1) countries with life expectancy stagnating at a high level, such as Denmark,

W. Apt, *Germany's New Security Demographics: Military Recruitment*
in the Era of Population Aging, Demographic Research Monographs,
DOI 10.1007/978-94-007-6964-9_3, © Springer Science+Business Media Dordrecht 2014

Japan, France and Poland; (2) countries with rapidly increasing life expectancy, including Algeria, Chile, Egypt, India, Libya, Mali, Morocco, Mexico, Tunisia and Senegal; and (3) countries with a decreasing life expectancy in recent years, such as Botswana, Iraq, Liberia, North Korea, Rwanda, Russia, Zimbabwe and South Africa. Until 2050, the UN expects life expectancy to increase from 78.0 years in 2010 to 83.3 years in more developed regions and from 67.5 to 75.1 years in less developed regions. In the least developed countries, the projections assume a respective increase from 58.8 to 70.2 years (UN 2011).

As the second major factor driving demographic trends, global fertility rates are in general decline. Since the 1970s the total fertility rate, i.e. the average number of children that a woman would bear if the current fertility regime remained the same over her lifetime, declined from 4.47 to 2.52 in 2010 children per woman at the global level. The United Nations expects that this rate will continue to decline (UN 2011). Yet such aggregated figures mask the existing heterogeneity in regional fertility developments.

In most industrialized countries, the fertility decline from above the "replacement level" of 2.1 births per woman set in around the year 1965 (and in the early 1970s in Germany). It can be explained by the large-scale social and cultural change and various strongly interrelated factors, such as the continued secularization, educational expansion, female emancipation, and individuation along with changed priorities about individual fulfillment, childbearing and family life (Lesthaeghe and Surkyn 1988; Lesthaeghe 1983; van de Kaa 1987).

As a result of these paradigm shifts, fertility levels in all of the 45 industrialized countries of the world are now below the replacement level (UN 2007: 10). Among them are countries with a *moderately low* fertility rate between 1.6 and 1.9 children per woman (including Australia, Belgium, China, Denmark, Finland, Netherlands, Norway, Sweden, United Kingdom) and other countries with a *very low* fertility rate of below 1.5 children (Austria, Bulgaria, Croatia, Germany, Greece, Italy, Japan, Poland, Romania, Spain, Hungary, all successor states of the Soviet Union) (EC 2007: 12).

Meanwhile, fertility levels in less developed world regions have decreased from 5.4 in the early 1970s to 2.7 in 2010. This fertility reduction is mainly attributable to the decline in childhood mortality, economic and social development, technological change that reduces the utility of child labor in home production, along with the establishment of capital or insurance markets that lead to a reassessment of children as a "security investment" and a paradigmatic shift from the quantity towards the quality of children (Cain 1983: 693ff.; Becker 1981: 110ff.). As a result, the number of developing countries with fertility below replacement level is increasing. Already 28 of the total 73 countries with below-replacement fertility are located in less developed regions of the world. They include Algeria, China, Egypt, Iran, Mexico, Thailand and Tunisia, which all experienced a rapid fertility decline in the past few decades (UN 2007: 9).

In the meantime, fertility levels are still above the replacement level in 122 less developed countries. Among them, there are 27 countries, where total fertility is still greater than 5 children per woman. Included are also 14 of the least developed countries,[1] where fertility levels still range above 6 children per woman. Although their fertility rates will decline until 2050, none of them is projected to reach the replacement level of 2.1 children. Hence, all of these countries will experience significant population growth (UN 2007: 10).

With regard to the third major demographic variable of migration, it is estimated that nearly 200 million people, i.e. about 3 % of the world population, live temporarily or permanently outside their country of their birth (IOM 2008a: 16). The number of international migrants has doubled in the past 30 years, whereas mobility among industrialized countries was particularly high (UN 2006).

Yet a significant share of today's transnational movements takes place within developing countries. According to estimates of Ratha and Shaw (2007: 6), about 73.9 million of the total of 155.8 million migrants that originate from developing countries also reside in other developing countries. In addition, such South-South migration is "overwhelmingly intraregional" (Ratha and Shaw 2007: 7). For example, 10.1 million migrants out of a total of 14.5 million migrants in Sub-Saharan Africa stay within the region. Similarly, Asian migration has evolved into an increasingly intraregional phenomenon (IOM 2008b: 439). Internal displacements or transborder refugee movements owing to natural disasters or violent conflicts are major migration issues in Africa and Asia (IOM 2008b: 407–416/439–452).

Much of this migration is absorbed by cities. The current urbanization process is characterized by two main trends: Firstly, megacities with populations of ten million people or more become larger and more numerous, especially in less developed regions of the world. Secondly, smaller urban settlements with fewer than 500,000 inhabitants absorb the lion's share of population growth (UN 2004: 3). In the foreseeable future, almost all of the world population growth will stem from urban population growth in the less developed and least developed countries in the world (UN 2008: 4).

In Europe, countries in Eastern Europe have been major sources of migratory flows, in particular Bulgaria, Poland, Romania, Russia and the Ukraine (IOM 2008b: 455–458). Some receiving countries like the Czech Republic, Italy, Greece, Slovenia and Slovakia, only sustained positive population growth due to the inflow of migrants. In other countries such as Germany and Hungary, the population decline would have been much larger without immigration. In sum, the EU-25 region had a net gain of 1.8 million of international migration, which contributed to nearly 85 % of Europe's population growth in 2005 (Münz 2007: 2f.).

The existing diversity in global and regional patterns of mortality, fertility and migration, and the ensuing disparity in the population composition and population change have been subsumed under the term "demographic divide" (Kent and Haub 2005). Given the influence of demographic factors on the economic development, social cohesion, internal stability and foreign policy orientation of states, this demographic divide may entail significant security consequences. According to Sciubba

---

[1] They include Afghanistan, Angola, Burkina Faso, Burundi, Chad, the Democratic Republic of the Congo, Guinea-Bissau, Liberia, Mali, Niger, Sierra Leone, Somalia, East Timor and Uganda.

(2011: 12), one of the most important security-relevant trends in the future will be the "growing divide in age structure between the aging industrialized great powers and the youthful industrializing powers."

3.2   The Demographics of Discontent

In the Malthusian theory of population, the combination of continued population growth and limited resource availability is considered a risk to the welfare and security of a society. In his seminal research, Homer-Dixon (1991, 1994) posits that rapid population growth and resource scarcity may bring about resource wars, group identity conflicts, or civil insurgency, each with the potential to seriously harm the security interests of the developed world (Homer-Dixon 1994: 7). Similarly, Kaplan (1994: 54) argues: "The political and strategic impact of surging populations, spreading disease, deforestation and soil erosion, water depletion, air pollution, and, possibly, rising sea levels in critical, overcrowded regions [...] will be the core foreign-policy challenge from which most others will ultimately emanate."

In the general discussion about the possibility of future conflicts over resources, water scarcity is a frequently mentioned issue (e.g. Fischer and Heilig 1997; Haftendorn 2000; OECD 2005; Postel and Wolf 2001). The chain of reasoning leads from population growth and improved standards of living to increased demands for water and an overuse of water resources that, especially in the case of transnational claims of use, may lead to conflict (Toset et al. 2000: 971f.; Wöhlcke 1996: 40f.). It appears however that, thus far, few conflicts were solely motivated by the lack or destruction of natural resources (Biermann et al. 1998: 288). Similarly, Goldstone (2001: 43) argues: "No nations have ever gone to war strictly over access to water; nor are any likely to do in the future." Since the advent of modern age, only seven small-scale armed hostilities over water have been documented, while about 3,600 agreements have been signed to regulate the utilization of transnational water resources (Wolf 1999: 6). These statistics tend to confirm the notion of environmental scarcity as both a variable associated with state's risk of civil conflict, and as a driver of more effective cooperation and more efficient resource governance (Toset et al. 2000: 976; Goldstone 2001: 42).

Nevertheless, the salience of demographic and environmental pressures has also been evidenced by the emergence of some real conflicts. For example, the Israeli-Palestinian conflict has been exacerbated by tensions over water rights in the West Bank. Water also played a major role in the conflict between Senegal and Mauritania, and in other interstate disputes in Africa. Turkey and India have used the regulation of water resources as a means of political coercion against other regional powers such as Syria, Iraq, and Pakistan. Moreover, climate change and the ensuing drought have fueled conflict over water resources in the Horn of Africa, while deforestation escalated insurgency in the Philippines (Hoyt 2003: 219).

In the foreseeable future, sustained population growth will lead to a significant drop in per capita water supply. This shortage will become particularly challenging in regions, which, already today, have low rainfall and a high population density, as for instance in the Middle East, and North, East or South Africa. At present, arid and densely populated areas in North Africa (e.g. Morocco, Egypt and Sudan) and Central or Western Asia already suffer from water scarcity (Rijsberman 2006). By 2025, the number of African countries affected by water scarcity or water stress (1,000–1,700 m^3 water per capita/year) will increase from currently 14 to 25. It is projected that then nearly half of the African population will live in countries with water scarcity or water stress, and that in particular, the water-rich regions adjacent to the rivers Nile, Niger, Volta and Zambesi will bear an increased conflict risk (UNEP 1999).

Most countries burdened with rapid population growth and environmental scarcity exhibit a youthful age structure with a low median age and high proportions of young dependents. The existence of such a "youth bulge", typically defined as a very high share of youth between 15 and 24 years in the adult population of 15 years and above, has also been considered as an indicator of an increased potential for conflict (Urdal 2006; Wagschal et al. 2008). For example, Huntington (1998: 259ff.) argues that societies are particularly prone to conflict when the share of young people of 15–24 years reaches a "critical level" of 20 % of the total population. However, Urdal (2004, 2006, 2011), one of the leading experts on youth bulges and civil conflict, finds no evidence that youth bulges above a certain level automatically make countries especially prone to conflict. In a cross-national study using demographic and conflict data from the period 1950–2000, Urdal (2006) finds a linear and statistically significant association between youth bulges and the risk of conflict. For each percentage point increase in the share of youth of 15–24 years in the adult population of 15 years and above, the risk of conflict increases by 4 %. He also shows that for countries, which experience youth bulges of more than 35 % of the adult population, the risk of conflict is 150 % higher than in countries with an age structure similar to most developed countries. According to the most recent UN population data, mean value of youth shares among the developed countries ranged at 15 % in 2010. Figure 3.1 shows the geographical distribution of youth bulges in the year 2010 with a size categorization of "low", "medium", "high" and "extreme" derived from the above literature. Countries with a share of youth in the adult population are mainly situated in Latin America, North and South Africa, the Middle East and Asia. Those with an extreme ratio of youth are almost exclusively located in Sub-Saharan Africa.

However, a simple causal inference along the formula "youth bulge = violence + conflict" falls short of the many variables at play (Wagschal et al. 2008: 356). The mere existence of a large youth population is not in itself a negative factor in the development of any country, nor a good predictor of conflict. Related evidence can be derived from the bivariate relationship between the 2010 Global Peace Index and the share of youth in the adult population in the same year (cf. Wagschal et al. 2008). According to the definition, an increase of the Global Peace Index means a decrease in peacefulness. Figure 3.2 shows a medium correlation between both variables. The dispersion of values is noteworthy. When countries with a high or very high share of youth are concerned, some countries have a notably high conflict incidence (e.g. Somalia, Iraq, Afghanistan, Sudan, Chad,

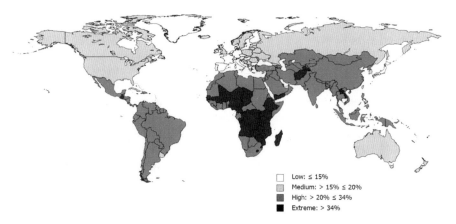

Fig. 3.1 Global distribution of youth bulges, 2010 (Source: Author's calculation and design. Data: UN (2010))

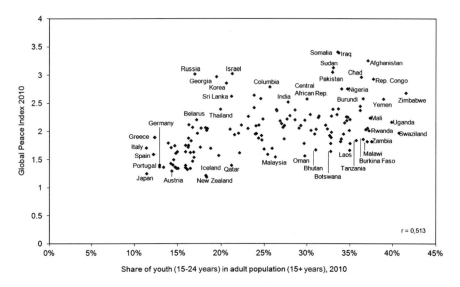

Fig. 3.2 Global Peace Index and share of youth population, 2010 (Source: Author's calculation and design. Data: Institute for Economics & Peace (2010) and UN (2010))

Israel and Columbia). Other countries display a lower level of conflict than would be normally expected by their youthful age structure alone (e.g. Zambia, Malawi, Burkina Faso, Tanzania, Laos, Botswana, Bhutan, Oman, and Malaysia). This is a strong sign that youthful age structures are a relevant variable in explaining conflict but that other conditions like political stability, the level of democracy, and the overall human development are equally, if not more important measures (cf. Nichiporuk 2000; Cincotta et al. 2003; Leahy et al. 2007). In this regard, Fuller

(2004: 5) writes "if societies lack the social infrastructure to integrate, employ and care for a growing population, the potential demographic benefits of a youthful population instead become a serious drain on the resources of the state and form a dangerously unstable element within society."

In the theory of youth bulges, the main emphasis lies on the population age structure. The focus rests on the size and development of a specific age group and the implications for the rest of society. It is assumed that the relative weight of young people constitutes a significant factor in the outbreak of political violence. The contributing factors are typically classified into opportunities and motives (Collier and Hoeffler 2004). In the center of the opportunity argument are structural conditions that provide the means to organize political unrest or incite a war against the government. These include financial means from "lootable" resources, such as diamonds in West Africa, timber in Cambodia, or cocaine in Columbia, weak government capacity, and low recruitment costs for soldiers (Urdal 2006). Under the motive perspective, the eruption of political violence is seen as a response to economic or political grievances, such as poverty, inequality or lack of democracy. These reduce the opportunity costs of combatants and turn rebel groups into a worthwhile alternative to earn a living (Collier and Hoeffler 2004). The emerging risk of conflict may be further exacerbated when, as is the case in most autocratic regimes, young males are excluded from other avenues of social status attainment (Wagschal et al. 2008).

The argument that a member of a large youth cohort has lower opportunity costs to engage in political conflict than an individual born into a relatively smaller cohort rests upon the Easterlin theory of cohort size (Urdal 2006: 610). Due to the large size of their cohort, youth are continuously faced with institutional bottlenecks and tend to have a lower relative income. The existing educational system and labor market are not prepared to absorb the bulge of new entrants. The potential for unrest worsens when a certain level of prosperity has been reached and the human capital of youth is relatively high, e.g. after an educational expansion.

Regardless of the political destabilization, social protests must not always have detrimental effects since they may become the vantage point for more representative governance and a more equitable distribution of wealth (Sciubba 2011: 23). The youth-led revolts during the "Arab Spring" of 2011 offer some illustrative empirical evidence (summarized in Apt 2011): Economic stagnation and unemployment were major motives for the outbreak of violence. In Egypt, Jordan, Libya and Tunisia, the economic situation had been deteriorating since 2008. Economic growth was negative in all of the countries. As a result, youth unemployment was very high and reached levels of about 30 % in Egypt and Tunisia, and 20 % in Libya.[2] In most Arab countries, the unemployed are faced with a significant risk of poverty. According to estimates of the United Nations, only 1 in 33 unemployed persons receive financial assistance from their government. Mostly affected were young people with a university education degree. In Morocco, 22 % of those with a tertiary degree were jobless. In Egypt and Tunisia, this still applied to 14 % of

---

[2] All data derived from World Bank (2011) unless otherwise stated.

youth. The International Monetary Fund estimated that a country like Egypt would have to create about 9.4 million new jobs to absorb all those that are currently or will be looking for a job in the near future. However, part of the problem was of the countries' own doing: In Egypt, guaranteed employment in the public sector for all university graduates led to a significant increase in enrollment rates. The respective share increased from 12 % in 1991 to 29 % in 2007. Similar increases could be observed in Algeria, Iran, and Tunisia.

Normally out-migration works as a safety valve in countries with a youthful age structure and when there is an oversupply of qualified workers (Urdal 2004: 624). However, the European Union tightened its refugee and migration policy towards North Africa in recent years. As a result, the migration balance in a country like Egypt increased from −536,000 in the year 2,000 to −291,000 in the year 2005. Hence, if fewer job seekers emigrate, there is an intensified competition for economic participation. The lack of employment also limited the social opportunities of youth. Combined with the rising cost of housing, it became a serious barrier to marriage. Observers speak of an "economic celibacy" or a "wait adulthood" since getting married or starting a family becomes hardly impossible without a normal job. As a consequence, less than 50 % of males in the Middle East are married by the age of 29 compared to 63 % in the late 1990s (Dhillon and Yousef 2009: 80f.). By and large, the Arab uprisings in early 2011 appear as the inevitable endpoint of a collective frustration among youth about their political, economic and social exclusion. With that, they also produce proof of an imperfect social modernization: The weak economy could not absorb the growing, well-educated population at working age. The overall increase in educational attainment increased expectations of prosperity and was conducive to the pursuit of immaterial values like democracy, or justice. In combination, the socio-economic hardships, economic stagnation and the lack of representative institutions lead to a crisis of legitimacy among the governing elites, which found its expression in widespread political protest. The driving force behind this movement were those mostly marginalized, namely youth.

In addition to the population age structure, Hudson and Den Boer (2002, 2004) rate the female-to-male sex ratio as an explanatory variable for violent conflict. The selection of male offspring, particularly in China and India, led to a significant and disproportionate surplus of males. In the security-relevant age group of 15–34 years, by 2020, this surplus is projected to range between 29 and 33 million males in China, and between 28 and 32 million in India (Hudson and Den Boer 2002). With their argument, the authors implicitly refer to the dichotomy of opportunities and motives. They are convinced that the recruitment of young, unattached males into the military or other paramilitary organizations is facilitated by the lack of family ties and alternative avenues towards social status attainment in these two family-centered societies in Asia.

Such imbalances in the female-to-male sex ratios may have implications for foreign policy and domestic affairs. The large military recruitment potential among young, unattached and nationalist males may boost claims to attain regional power status. Similarly, it is conceivable that an increasing number of young, less educated males from marginalized layers of society engage in criminal behaviors, either directly or indirectly linked to the prevalent sex ratio. These may include human trafficking, violence against women, or other more general criminal conduct as a way to cope with the social and economic marginalization (Hesketh and Xing 2006).

Organized crime, violent delinquency and civil unrest mostly occur in urban areas. In the past, issues like rising food prices, increasing food scarcity, and currency devaluation have created public uproars in such cities as New Delhi, Jakarta, Karachi, and several African cities. On various occasions, such urban-centered violence was further fuelled by smoldering ethnic or intra-group tensions, as witnessed between Hindus and Sikhs in New Delhi, militant Hindus and Muslims in Ayodhya (with the conflict later spreading to other Indian cities like Bombay, Calcutta, and New Delhi), Mohajirs and Pathans in Karachi, or Indonesians and ethnic Chinese in Jakarta (Brennan 1999: 16; Nichiporuk 2000: 52). According to Brennan (1999: 16), such eruptions of urban violence have "serious potential for destabilizing worldwide financial markets and destroying infrastructure, thereby impacting already fragile national economies, or igniting violence in entire geographical regions."

More latent than such momentary outbreaks of violence, however, are the destructive effects of large-scale poverty, poor infrastructure, inadequate sanitation conditions and environmental degradation on public health and labor productivity, in particular of low-income megacity residents, which may evolve into an obstacle to the development of urban economies in populous, fast-growing and less developed nations. Since conditions in the urban centers will shape the economic performance, social development and political stability on the aggregated national level, urban structural deficits may hamper the development process at large and sustain the segregation of the world economy (Brennan 1999: 19; Wöhlcke et al. 2004: 143).

By 2025, there will be 20 so-called megacities with populations of more than 10 million, including nine cities with more than 20 million inhabitants (Sciubba 2011: 110). However, the lion's share of urban growth will not take place in such megacities, but in small- to medium-size cities with fewer than 500,000 inhabitants (UNFPA 2007). Many of them are ill-equipped to absorb significant population growth as their governance capacity, resource endowments, and service sector are less developed than in the largest cities (Brennan 1999: 8). With the majority of the world population living in cities or denser settlements, and given the rapid urban growth in developing countries, the likelihood of armed conflict in urban environments increases (Nichiporuk 2000: 54). Past military interventions in the densely populated urban areas of Beirut, Sarajevo, Mogadishu, Grozny, Kabul, or Baghdad entailed significant costs in terms of time, personnel and materiel. The prospect of an increased frequency of urban warfare in the future appears unsettling to military planners given that most tactical superiority of Western allied forces, in the form of close air support, artillery, and armor, would be of little use in the "urban labyrinth of building, roadways and sewers" (Nichiporuk 2000: 54).

By way of conclusion, Krause (2006) suggests a helpful classification of less developed countries and their associated demographic risk factors. In the first group, he classifies countries, where population growth goes hand in hand with economic growth, and where the blessings of economic progress benefit all layers of society equally. The second group of countries consists of failing or failed states, which are extremely poor and continue to display extreme disparities in wealth and an unequal access to resources. The two groups of countries yield different consequences for international security. Countries in the first category may challenge the existing structure of global governance, such as the distribution of votes in international organizations like the United Nations Security Council. By means of their increased economic activity and societal modernization, these countries will also make use of more natural resources, and thereby exacerbate environmental scarcities. While the first group of countries will increasingly question the existing rules of global governance, trade and security, the second group of countries will likely be the stage of demography-induced conflict or instability. Possible consequences of these disruptions are diverse and detrimental to the development process: They include migration, food insecurity, poverty, the criminalization of politics and society, and environmental damage. In any case, they tend to increase the demand for Western military intervention.

3.3   The Demographic Variable in Foreign Policy

The study of population as a source of relative power in the international system has a long tradition. Already Thucydides' *Melian Dialogue* about the Athenian conquest of the island of Melos in 416 B.C. linked the rise and fall of nations to demographic trends. The underlying hypothesis has always been that population size translates into military power, which is again indispensable to attain great power status. In this vein, the well-respected international relations scholar Hans J. Morgenthau asserted that "no country can remain or become a first-rate power which does not belong to the more populous nations of the world" (cited in Jackson and Howe 2008: 80). Larger nations, by the virtue of their sheer size, possess larger national budgets that they can devote to military purposes, and they also have a larger pool of sophisticated human capital to boost the economy and develop cutting-edge technology (Weiner and Teitelbaum 2001: 15).

For the complex system of nations today, the causal linkage of population size and international significance seems too static and reductionist. Israel is an apt example of a country with a small population that fares well relative to more populous states (Weiner and Teitelbaum 2001). Therefore, state power does not only depend on demographic variables. It rather foots on the combination of political, organizational, economic and cultural capabilities (Tellis et al. 2003). This is in line with the writings of Nazli Choucri about the interplay of population and violence, in which she explains that "only through the mediated influence of technology, social organization, capital and military equipment, and political structure, among other variables, does population become an important consideration" (Choucri 1974 cited in Krebs and Levy 2001: 65).

In view of this indirect demographic effect, gradual population decline does not have to equal a decline in national power and prosperity (Tellis et al. 2003). However, seeing that a long process of population aging precedes the actual drop in population size, the transition from a larger to a smaller population will be a difficult one for many nations (Longman 2004). Here the concept of the "effective population" comes into play, which refers to the share of the population that is instrumental in furthering national goals (Organski et al. 1972 cited in Krebs and Levy 2001: 65). The concept represents a more meaningful alternative to the static definition of national power on the basis of population size and has been coined as the "human capital paradigm of population and national power" (Nichiporuk 2000: 8). It considers a population's skill level and age structure. Hence, not the size of the total population is pivotal for nations' relative power, but the relative changes in their effective population. In economic terms, the effective population equals the share of people that actively participate in the labor force (Krebs and Levy 2001: 65).

With population aging, the size of this effective population declines. The French demographer Alfred Sauvy characterizes an aging society as one consisting of "old people, living in old houses, ruminating about old ideas" (cited in Weiner and Teitelbaum 2001: 31). In his "population senescence hypothesis", he suggests that an older society will be afflicted by "a lack of political vigor, creativity and ambition, economic vitality, social dynamism, and military prowess" (Weiner and Teitelbaum 2001: 31). The notion that an older age structure is attended by changes in social mood and political behavior has been discussed for a long time. For example, in 1949, the British Royal Commission on Population Report (cited in Vincent 2003a: 1) noted: "Older people excel in experience, patience, in wisdom and breadth of view; the young are noted for energy, enterprise, enthusiasm, the capacity to learn new things, to adapt themselves, to innovate. It thus seems possible that a society in which the proportion of young people is diminishing will become dangerously unprogressive, falling behind other communities not only in technical efficiency and economic welfare but in intellectual and artistic achievement as well."

In terms of states' foreign policy orientation and military capabilities, population aging and potential changes in social mood may have three major implications: Firstly, elderly-dominated electorates may shift political direction towards a more inward-looking agenda-setting and an all-important attention to securing domestic welfare and internal stability at the expense of foreign policy objectives and collective security commitments. Secondly, the numerical increase in pensioners will inevitably translate into higher public expenditures for age-related items such as statutory pension or healthcare payments, whereas the decline in the population share at working age will yield a reduced government revenue from income tax, which, in combination, increases the risk of so-called "guns-versus-butter-tradeoffs". Thirdly, smaller family sizes may leave parents, especially those of only children, less affirming of the military profession as an occupational choice for their scarce offspring. Although uncertain, given the lack of predictability of social change or future political decision-making, each of these topics is discussed against the background of the existing literature. The objective is to outline a few potential demography-induced constraints of foreign policy formation.

3.3.1 Politics of Aging

Foreign policy may be constrained by changes in public preferences. Despite "significant inertia" (Dalgaard-Nielsen 2006: 10), state preferences are not entirely static but are subject to change under severe social pressure. Similarly, Choucri and North (1972: 94f.) argue that the latitude of foreign policy decision-makers may be potentially constrained by real budgetary expenses and political costs of interventions far away from home. While the former results from equipping, manning and managing military operations, the latter may evolve from the dissatisfaction of the domestic constituency with the government's foreign policy agenda. The level of such political costs tends to increase with the duration, and hence the real costs of the engagement, and the distance of the operational area from home.

It becomes clear that states do not adhere to a given conception and practice of security. Rather, policymakers pursue combinations of different public goods that represent the will of powerful domestic groups (Moravcsik 1997: 519f.). This becomes even more problematic in view of what McGuire (2000: 22) calls an unconcealable "vagueness of the connections between [national] self-interest and remote conflict", which may, in combination with budget squeezes, fuel uneasiness among the publics of industrialized nations to shoulder the burdens of international crisis intervention. It therefore appears that, in the foreseeable future, increasing social pressure may arise from the graying of the industrialized countries' electorates and potential distribution conflicts between different age groups.

Demography-induced shifts in public preferences and voting power are uncertain and difficult to predict empirically. However, there are several, mostly U.S.-based, sources that argue that senior power is on the rise across Europe and identify electoral politics as a key challenge to future foreign policy. For example, with reference to the social and political turmoil caused by proposals to modestly cut elderly welfare payments in Italy, the Netherlands, France, Greece and the U.S. during the past decade, Haas (2007: 123) illustrates the difficulty of governments to adapt social policy to the new demographic realities. For politicians, this resistance becomes even more relevant in view of the sheer numerical weight of the elderly at the voting booth and their relatively higher voter participation. In a similar vein, Jackson and Howe (2008: 128) argue that the elderly vote is effectively greater than the relative size of the pensioner population would suggest given that workers at advanced age already anticipate their retirement well before the actual retirement and adapt their voting behavior accordingly.

Against this background, a report by the CIA (2001: 85) predicts that by 2030 at the latest, most European nations will equal "demographically challenged, fiscally starving neutrals who maneuver to avoid expensive international entanglements" given the political power of elder-dominated electorates, which will be "more risk averse, shunning decisive confrontations abroad in favor of ad hoc settlements." Similarly, Jackson and Howe (2008: 131) suggest that the absolute increase in the elderly population across the industrialized world will bind public resources, lock in current spending priorities, and block potential resource shifts to new policy priorities, whether in the internal or external environment of states. However, in the realm of foreign policy, demographic aging could result in a "more cautious and risk-averse mood" of the society at large.

Other research in Political Economy shows that the population age structure defines the relative size of voting groups at each age and thereby predetermines the government's ability to enforce reforms in age-specific issue areas (Lee and Edwards 2001: 191; Poterba 1997: 51). In this connection, it has been argued that changes in the age structure entail a relative reallocation of aggregate public resources away from the younger population towards the middle- and old-age population (Teitelbaum and Winter 1985: 108).

Descriptions or interpretations of the political orientation and voting behavior of the elderly population bear distinct differences and largely depend on the national context from which conclusions are drawn. For example, coverage in the United States is largely dominated by the "myth of the grey vote", which depicts "seniors as a large, homogenous, self-interested bloc" (Holladay and Coombs 2004: 385). From this perspective, the imminence of a senior voting bloc and resource-driven conflicts between generations appear like a realistic scenario. The evidence and framing in the European-grounded literature seems a little more nuanced. Although the "time bomb" image is likewise used (e.g. McKie 2004), the fear of the elderly as a political power bloc seems less prevalent.

For example, Vincent (2003: 10) seriously doubts the political significance of the elderly population. His main criticism targets the concept of a generation as a (homogenous) social category or distinctive identity that would serve the elderly population as the main reference point for political coalition-building or the pursuit of economic well-being. Rather, Vincent (2003: 12) argues that the elderly elector- ate does not engage in tactical voting and is not more driven by individual benefit maximization considerations than other age groups. Similarly, Hamil-Luker (2001: 386) has empirically shown that age is not a strong predictor of public preferences in the federal budget allocation process. She highlights the combination of socio- demographic cleavages like sex, race, educational attainment, income, subjective class location, and political affiliation, which altogether reduce the potential for generational conflict. In the same way, Bernhard and Phillips (2000: 42) explain that the changes in the socio-demographic makeup of today's industrialized societies, largely due to longer lives, lower fertility, higher divorce rates, or higher educational attainment, have influenced the elderly in very different ways. Depending on their individual set of circumstances, they are heterogeneous in their dependency on social welfare and their propensity to become politically active. Also, Ryder (1965: 858) argued that in later life, cohort identity or generational affiliation become less important since age becomes an increasingly imprecise measure of an individual's social characteristics. Thus, people at older ages vary in their socio-economic back- grounds and human capital endowments, such as education and health, and there- fore have specific personal needs. This disparity in attitudes and preferences reduces their political group power. In this vein, Vincent (2003: 12, 2005: 592) empha- sizes the complexity of old age politics, and the fact that the elderly do not consti- tute an undifferentiated category. While age seems to be positively correlated with political conservatism, older people also exhibit a significant diversity in political views (Vincent 2005: 593).

Political divergence and social divergence are closely related. The latter mainly arises from differences in social class, retirement benefit status, savings, or health status. It undermines political cohesion and prevents the elderly from emerging as a politically effective force. It is difficult for them to agree upon and realize common goals. As a consequence, Vincent (2003: 11f.) concludes that the elderly do not seem to act collectively, vote together or use their demographic weight to dominate politics per se. Based on comparisons between the United Kingdom, France, the Netherlands and the United States, he finds that concern over the growing electoral power of the elderly and potential repercussions for intergenerational equity mainly stems from the United States, which, in fact, possesses the relatively weakest social welfare system and the politically most effective organization of older people (Vincent 2003: 13). Similarly, Hamil-Luker (2001: 387f.) provides a persuasive critique of claims about an impending intergenerational conflict. According to her statistical analyses of public opinion data on old-age related government spending, the widespread concerns rather stem from the existing mix of public policies than a society's demographic composition. Thus, at least, in the European context, the idea of senior power appears to be overstated and mainly arises from the confusion of the electoral power of older people with the popularity of the welfare state (Vincent 2003: 12). Moreover, the social construction and interpretation of generations foster the generation-based rationalization of the political process (Vincent 2005: 583).

Nevertheless, Vincent (2003: 2) acknowledges the possibility that people of approximately the same age develop a sense of generational identity on the basis of common experiences. Such a perceived distinctiveness could also aid the formation of generation-specific political preferences, moral concepts and expectations of entitlement. Such perceptions would influence generations' political orientation and voting behavior, while providing a basis for collectivizing efforts towards economic interests as regards property rights, occupational positions, social benefits and pensions. World War II represents one historical circumstance that has formed a sense of collectivity and common identity. As a uniquely grueling experience of a lifetime, the group experience of war shaped the values, attitudes, and national identity of an entire generation. According to Vincent (2005: 592–595), those sentiments are expressed in the comparatively high voter participation among the elderly, and in the case of the United Kingdom, are reflected by the large number of "Eurosceptics" among older age groups that tend to interpret international relations, in particular with France or Germany, through the lens of their war memories.[3]

There are some examples of governments that responded to the pressure of an increasingly powerful elderly electorate by adjusting social policy to old-age needs and allocating a disproportionate share of public resources to the pensioner

[3] In turn, under the assumption that the collective experience of war makes people more prone to a rather pacifist foreign policy strategy, potentially centered on soft power, it remains to be seen how societal preferences among the baby boomer generation, i.e. the elderly electorate of tomorrow, will evolve in absence of such a generation-specific security threat.

population. For example, in the face of large-scale social tensions, the Russian government under Putin felt impelled to raise benefits for 40 million pensioners by more than 8 % in 2005 (Haas 2007: 124). Similarly, the announcement of the Chinese government in 2002 about a tightening of the criteria for statutory pension payments caused such civil unrest that, in 2004, the government committed itself to social security as a cornerstone of long-term social and political stability and announced a significant increase in the National Social Security Fund by 2010 (Haas 2007: 124). While this does not mean that states are generally unable to curtail social benefits in order to have adequate budgetary room for discretionary spending like defense, it does provide insight into the dominant societal preferences. In this context, taking a look at Japan, the "laboratory example" of demographic change given its already well advanced aging process, seems worthwhile. Reflecting upon public opinion, Agakimi (2006) notes: "Japanese society is rapidly ageing, and the population is shrinking. What the people care about are social security, medical care, neighborhood safety and the like. An aggressive nationalism of geriatrics would indeed be a historic feat."

In Germany, the median age is projected to increase from 43 years in 2008 to 52 years by the mid-2040s. Hence, between 2045 and 2060, about half of the German population will be older than 52 years (Federal Statistical Office 2009: 16). Accordingly, the relative share of the elderly population will increase. When the ages 60 years or older are concerned, their share of the total population will rise from 25.6 % in 2008 to 36.8 % in 2030 and above 40 % by 2050 (Federal Statistical Office 2009: 39f.). Given their demographic weight and high political participation, the elderly in Germany may possess significant political group power in the future. Against this background, Seitz (2007: 140) expects that the aging of the electorate will boost an age-conscious political agenda-setting and favor political parties with a strong orientation towards social policy and welfare provision. Other observers assume that future policymaking in Germany will remain resistant to demographic change and continue to target all policy areas in balanced way (Streeck 2007).

However, anecdotal evidence tells otherwise. During the German national election campaign in 2009 and right at the onset of a global economic crisis, some social democrats issued a "guarantee" that the pension level would not regress irrespective of the general state of the economy or possible salary cuts among the working population. When this proposal was passionately rejected by the Christian-Liberal parties, the president of an old-age lobbying group frankly stated: "Those who question the retirement pension guarantee have to be prepared to get punished at the national election" (Wolf 2009). This occurrence indicates of the potential political power of a growing elderly electorate.

Although younger age groups in Germany have expressed a greater willingness than the older population to pay for defense (Bulmahn et al. 2008: 129), the origin of this age-specific divide remains unclear. Does it stem from socio-economic differences, from the collective remembrance of World War II among the elderly that led to a pacifist attitude among many, or from domestic welfare considerations and a preoccupation with the securing of private goods like retirement benefits?

Table 3.1 Projected government benefits to the elderly population, % GDP

	Public pensions			Health benefits			Total		
	2005	2030	2050	2005	2030	2050	2005	2030	2050
United States	6.1	10.4	11.0	3.2	7.6	10.4	9.3	17.9	21.4
Canada	4.4	8.3	9.7	3.1	5.6	6.9	7.5	13.9	16.6
United Kingdom	6.6	9.4	11.2	2.7	5.5	8.7	9.3	14.9	19.9
Germany	11.7	18.4	22.6	3.4	4.8	6.8	15.1	23.2	29.3
France	12.8	19.0	22.1	3.5	6.6	9.4	16.3	25.7	31.5
Italy	14.2	19.3	27.6	2.8	4.0	6.0	17.0	23.2	33.6
Japan	8.7	14.3	20.2	3.4	4.4	6.0	12.0	18.6	26.2

Source: Jackson and Howe (2008: 65)

While it may appear that today's working population and, hence, tomorrow's retirees will be more willing to pay for defense, two objections should be made. Firstly, available survey results show that people evaluate foreign policy engagements against current economic trends and social conditions (Bulmahn 2008: 14). Secondly, in the event of diminishing federal tax revenues, a large share of the general population would vote for budget tradeoffs at the expense of defense spending (Bulmahn 2008: 36). These trends support the statement made by Smith (2005: 499) that "public opinion is dynamic rather than static". It depends on changes in the broader social, economic and political environment. Given the large degree of uncertainty associated with public opinion formation, predictions about future societal preferences are merely speculative. However, the sheer possibility of an ongoing socio-political devaluation of foreign affairs and military issues in an aging society like Germany cannot be eliminated.

3.3.2 Budgetary Tradeoffs

Following basic fiscal principles, military spending has to be financed through the public sector. The potentially adverse impact of demographic change on defense spending arises from declining tax collections from a dwindling workforce and a simultaneous increase in age-related spending. In all likelihood, demographic changes will pose serious fiscal challenges in the future. According to estimates of Jackson and Howe (2008: 64f.), the cost of public pensions and health benefits for the elderly population will increase in all G-7 countries, albeit varying in magnitude due to national differences in demographic outlook and the generosity of benefit systems. Against this background, the Anglo-Saxon countries are expected to record only modest increases in old-age government spending (Table 3.1). Germany's annual payments to the elderly are projected to rise from a total of 15.1 % in 2005 to 23.2 % in 2030 and, regardless of increasing forecast uncertainty, to 29.3 % in 2050. The situation in France and Italy may become even more staggering since old-age related government spending in both countries is projected to exceed 30 %.

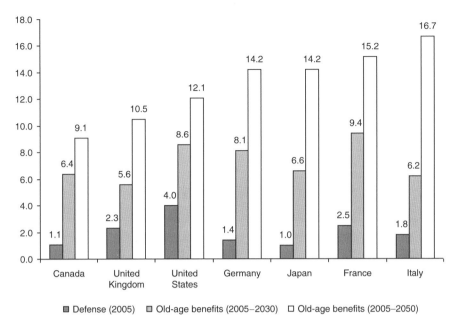

Fig. 3.3 Defense spending vs. old-age benefits, % GDP, 2005–2050 (Source: Jackson and Howe (2008: 89))

In order to shed light on the magnitude of the looming fiscal challenge posed by demographic change, Jackson and Howe (2008: 89) provide an illustrative sample calculation for future old-age related spending compared to current defense outlays among G-7 countries. All countries will face significant increases in old-age benefits. They will, in fact, be many times higher than the amount that these states currently spend on defense. For example, Germany will be faced with an increase of its annual spending on old-age benefits by more than eight times of what it now spends on defense (see Fig. 3.3).

In order to accommodate this substantial increase in age-related welfare spending, governments have four principal compensatory measures: raise taxes, increase deficit spending, reduce benefits (including the raise of the statutory retirement age) or cut spending in other discretionary spending areas where funding is not mandated by law, such as defense, education or infrastructure (Haas 2007: 121). However, it appears most likely that the governments of aging industrialized countries like Germany will curtail defense spending in order to pay for the new expenditures required by demographic aging (Jackson and Howe 2008: 90). The reasons are multifaceted and include the limits of tax efficiency, the heavy reliance of pensioners on government support, the political unacceptability of elderly benefits cuts and pensioners' electoral power, the binding character of the European Union's Stability and Growth Pact that mandates national fiscal policy, and the relatively low priority that domestic constituencies seem to assign to foreign policy and international security issues.

The increase in domestic welfare spending does not deplete the federal budget completely. National leaders rarely ever lack the de facto capability to deploy military force. Yet, military effectiveness largely depends on resource availability. In particular, when the provision for domestic demands is at risk, foreign policy engagements are particularly difficult to justify (Clark and Hart 2003: 63f.). Hence, national leaders may be limited by the electoral threat to authorize foreign policies that may be costly in terms of material and human resources (Clark 2001: 641).

On the assumption of rational choice, policymakers have an interest to retain office and an incentive to implement policies that serve their reelection (Clark and Hart 2003: 60). Distributive or redistributive policies, which provide economic security and social welfare for core constituencies, have been found to be one of the most effective strategies for this endeavor (Clark 2001: 643). Such private goods or policies, which approximate private goods, grant benefits to certain parts of the population based on legally defined allowance criteria, such as age limits or financial neediness, and are denied to the general population. Meanwhile, in view of their non-rivalry in consumption and non-excludability (Kapstein 1992: 41), public goods like national security are "largely ineffective" to attract support of constituencies and increase reelection chances (Clark 2001: 643).

The traditional resource conflict of "guns versus butter" has been widely discussed in the defense economic literature. Tradeoffs among budgetary expenditures occur when the spending level for one category affects the amount of money allocated to another category, such that spending for the first category comes at the expense of the second (Berry and Lowery 1990: 674). During the Cold War, a widely held belief was that defense requirements would lead to the retrenchment of overall civilian spending. However, related empirical research only offered inconsistent and oftentimes conflicting findings. From a comparative analysis of the results, Domke et al. (1983: 20) observed a general tendency of "industrialized democracies to avoid sacrificing social needs in the face of security requirements" during the Cold War. Newer research by Clark and Hart (2003: 64) confirms this notion by arguing that a reduction in social welfare spending "to protect other types of policy is politically difficult (if not impossible)".

By implication, as resources available for foreign policy become scarcer, national leaders will choose the use of force in a highly selective and purposeful manner (Clark and Hart 2003: 63). Similarly, Sprout and Sprout (1968: 674ff.) show that social welfare traditionally enjoys a higher budgetary priority than military spending in a major power like the United Kingdom, which they ascribe to the "soft" character of military commitments in the political process. Domestically, this flexibility of military spending forcedly arises from the binding character of other programs. Meanwhile, internationally, the softness of military commitments results from the non-enforceability of alliance agreements, which gives rise to the so-called "free rider problem" that has also been a prominent issue in post-Cold War alliance formation (Kapstein 1992: 162).

Other concomitant factors in the social, economic and political sphere that may adversely affect military spending in post-industrial societies include (1) the educational expansion and increased political activity of the general public, which entails higher social welfare demands and increased political leverage; (2) the priority of subsidies for advancing industrial technology and methods of production in order to secure the international competitiveness of the domestic economy; and (3) the perceived infeasibility to achieve the social and economic goals mentioned above other than with cuts in defense spending based on "a premise rarely made explicit in public – that it is politically inexpedient to dampen private spending severely or to cut back governmental spending for the social services" (Sprout and Sprout 1968: 672–678).

These domestic issues surely have geopolitical implications. According to Haas (2007: 128f.), who classifies the magnitude of the aging challenge in terms of eight demographic, economic and fiscal criteria to make predictions about the future ability of states to finance military implications, the United States will strategically benefit from the demographic changes underway elsewhere (Haas 2012). These will bind public resources that could have been used for an active foreign policy agenda, potentially challenging American interests (Haas 2007: 144). However, the predicted primacy comes at an expense. The capacity of allies to provide for economic and military assistance in crisis-ridden areas will be fiscally and politically constrained.

3.3.3 Casualty Aversion

The limited success of humanitarian interventions in Somalia, Rwanda and the former Yugoslavia during the 1990s has been, among other things, attributed to a phenomenon called casualty aversion (van der Meulen and Soeters 2005: 483; Ben-Ari 2005: 652; Luttwak 1994: 24). Prominent display of this public and political unwillingness to risk soldiers' lives occurred subsequent to the death of 18 American soldiers in Mogadishu in October 1993 and the killing of 10 Belgian paratroopers in April 1994, which resulted in the removal of U.S. troops from Somalia and the abrupt removal of the Belgian contingent from Rwanda (Smith 2005: 496; Manigart 2005: 560). Moreover, the universal shift of Western military forces towards campaigns of precise, limited destruction from great distances seem to confirm that concern over casualties has grown stronger and more influential in recent years (Manigart 2005: 560; Smith 2005: 506). This use of precise-guided weapon technology also created the public illusion of zero-casualty wars (Ben-Ari 2005: 653).

One of the most influential proponents of the casualty aversion paradigm has been the American military strategist and historian Edward N. Luttwak (1994, 1995). He argues that family size is an important variable in the low tolerance for casualties among Western societies. Luttwak's most general proposition is that changes in the demographic composition of Western nations have created a high level of casualty aversion and influence decision-making about the participation in military interventions. Luttwak (1994: 25) compares the average family with two children in the contemporary industrialized world with the historic family demography when families were large and fertility was high to compensate for the frequent death of children through disease, malnutrition or war. From this, he concludes that today's parents must be relatively more opposed to military service given that all children of the family are expected to survive into adulthood, "and each of whom represents a larger share of the family's emotional economy" (Luttwak 1994: 25, 1995: 115). The political power of combat losses derives from their societal diffusion; they are not confined to the parents and relatives of soldiers on active duty but shared throughout all levels of society opening out into "an extreme reluctance to impose a possible sacrifice that has become so much greater than it was when national populations were perhaps much smaller but families were much larger" (Luttwak 1994: 25). In this connection, Luttwak (1994: 27) maintains that the new family demography makes all advanced societies, most prominently France, Germany, the United Kingdom, Japan, Russia, and the United States, relatively averse to war. However, the casualty effect does not seem to be limited to industrial democracies. The prudent strategy of the Soviet military in the Soviet-Afghan War may be equally seen as proof of public pressure (Nichiporuk 2000: 20; Luttwak 1994: 24). Hence, the sweeping demographic shift in China due to the one-child policy may make the country equally averse to military confrontation (Smith 2005: 501).

Three key implications arise from this fear of casualties, presumably induced by demographic change. Firstly, industrialized nations may be faced with a diminishing power of coercion as potential belligerents observe the societal and political resistance to put soldiers at risk (Luttwak 1994: 23). Secondly, the public fear of casualties may prompt Western governments to select their battles even more carefully and constrain the range of available foreign policy options (Luttwak 1994: 28). In this connection, protracted decision-making on military interventions may aid the escalation of a security crisis and reduce the chances of ad-hoc settlements. Thirdly, casualty aversion may hinder mission success by keeping soldiers away from the true theater of war (Smith 2005: 488) and due to an overly reliance on far-distance weapon technology. Indeed, experience has shown that Western nations increasingly shift to alternative force categories, namely maritime and air power, in order to lower the risk of suffering casualties although the effective containment of crises and restoration of order still requires large-scale infantry, albeit flexible and highly mechanized (Luttwak 1995: 110; van der Meulen and Soeters 2005: 483). The result of such casualty-avoiding interventions would be "partial, circumscribed and often slow" (Luttwak 1995: 115).

Potential repercussions for military effectiveness have also been discussed in terms of a phenomenon called "structural disarmament".[4] It describes a situation, in which the fear of losing highly trained personnel inhibits operational readiness. One incident described by Boëne (2003: 171) illustrates the context: During the Gulf War, young French military officers were called for support in the Saudi desert. However, some battalion commanders refused to send off graduates from the prestigious Saint-Cyr Military Academy and explained that their comprehensive

---

[4] The issue of "structural disarmament" is equally at work when commanders are reluctant to use expensive weaponry in theaters of war perceived as marginal to the overall outcome of an intervention.

and expensive training made them "too precious to be wasted in [...] a sideshow." Similarly, subsequent to the traumatic experience in Rwanda, a special Belgian parliamentary commission recommended the immediate withdrawal and urged the military to "guarantee the maximum safety of military personnel" in future operations (Manigart 2005: 560).

The rationalization of politics and strategic calculus on the basis of demographic change certainly has some appeal. However, the true impact of diminishing family size is difficult to assess (Smith 2005: 501). It is beyond doubt that governments and military authorities are liable to some level of suffered or potential casualties in their decision-making about military operations (Kümmel and Leonhard 2005: 523). However, governments purposively conceal their true concern over casualties in military operations seeing that "enemies will be encouraged, allies discouraged" (Smith 2005: 489). This point is particularly relevant for the increased engagement of Western militaries in contemporary asymmetric conflicts. In this connection, Van Creveld (2008: 227) notes: "for the strong every [casualty] ... becomes one more reason to end the struggle. For the weak, it is one more reason to continue until victory is won." Despite the critical factor of human losses, the strategic underpinning of military commitments is more complex than the popular demography-based interpretation of casualty aversion. It commonly rests upon a rational, case-specific cost-benefit calculation and a collective consensus process among all relevant decision-makers to find the most suitable policy option (Smith 2005: 487, 496).

Those policy options selected typically vary across countries and depend on the society-specific concern about the casualty factor. In this vein, Smith (2005: 500) considers four social broad trends that foster the casualty avoiding mentality: (1) the simple facts of demographic change, most prominently the shift towards an older age structure and a declining proportion of children; (2) the decreasing reconcilability of the hardships during military service and the individualized lifestyles of contemporary youth; (3) the growing concern over the selectivity of military service and inequitable sharing of the burdens of war among different social groups; and (4) changing societal expectations about the nature of conflict and shifting attitudes about the use of organized violence, especially the utility, morality and justification for the use of military power.

In effect, "casualty aversion varies from conflict to conflict, nation to nation, and era to era, and it depends on perceptions and beliefs that may or may not [always] be based on fact" (Smith 2005: 507). Consistent with this proposition were (1) the demand of the American public in late 1983 to continue the U.S. military intervention in the Lebanese Civil War, rather than abandon it, in spite of the killing of 241 U.S. Marines in a Beirut barracks bombing (Smith 2005: 496); and (2) the general tolerance of casualties during the military operations in Afghanistan and Iraq, which, at least, for a long time, ranged on a level that was much higher than the conventional wisdom of the 1990s and the demography-centered paradigm by Luttwak would have predicted (van der Meulen and Soeters 2005: 483f.). On the other hand, there was a greater prominence of the casualty factor during many conflicts during the 1990s, which then ranked low on the scale of national interests. In terms of an underlying cause, Smith (2005: 493) argues that these conflicts were perceived as marginal to the security of Western states, which made the risk of the deployment of many soldiers politically unjustifiable.

It appears that the contextualization of military operations plays a decisive role in the legitimization and explanation of casualties. For example, in the more recent past, the German government framed peacekeeping in terms of worthy cosmopolitan and humanitarian objectives such as peace, democracy, freedom and human rights. It was also conveyed that Germany would have to contribute its fair share to the safeguard of international security (Kümmel and Leonhard 2005: 530; Ben-Ari 2005: 656). Still, the obvious impossibility of zero-casualty wars complicates the recruitment of quality personnel in today's risk averse societies (Manigart 2005: 559). Moreover, the post-heroic character of Western societies makes honorable causes or appeal to patriotism and duty less effective in producing legitimacy of casualties or attracting youth into the military (Smith 2005: 491). In this context, Luttwak (1995: 115f.) poses the fundamental, yet somewhat unsettling, question: "How, therefore, can armed forces, staffed by professional, salaried, pensioned, and career minded military personnel who belong to a nation intolerant of casualties [and fiscally strained by demographic change], cope with aggressors inflamed by nationalism or religious fanaticism [that possess a sizeable demographic recruitment base]?".

Military deaths, particularly during peacekeeping operations, significantly challenge the viability of the military organization and the framing of contemporary missions (Ben-Ari 2005: 654). In this context, two issue areas stand out. Their significance can be exemplified by the above-mentioned assassination of ten Belgian paratroopers by Hutu government troops while trying to protect the Rwandan prime minister. First and foremost, the incident made aware that peacekeeping operations, just as offensive military campaigns, can be dangerous – even in the case of low intensity operations in a relatively stable environment, which are conducted by well-equipped, internationally mandated military forces with the best intentions to restore political and societal stability (Manigart 2005: 559). Secondly, the contemporary military role set of peacekeeping entails a new set of soldierly risks that, in the event of a soldier's death, are difficult to conciliate with the "cultural script – that is [...] the widely accepted social scenario – of a good military death" (Ben-Ari 2005: 655). In this context, Smith (2005: 506) writes fittingly: "Dying heroically for one's country is one thing; dying from a bullet in the back at the hands of a bandit, or for the sake of another's corrupt and violent society in the course of law enforcement is another." As a consequence, there is little public approval of soldiers' deployment in support of causes that are perceived as largely obscure and unclear in terms of mission objectives, designated end points, financial cost, soldierly risks, allied support and the chances of success (Smith 2005: 495f.). Hence, the withdrawal must not always be a proof of casualty aversion but can also be the acknowledgement that "the operation was a mess", which turned out into a "losing cause" (Smith 2005: 496).

Albeit demographic change, it is unlikely that decision-makers will ever lack the material capability and human resources to project military strength. Rather, societal sensitivity to battlefield fatalities may render contemporary industrialized societies, as Luttwak (1994: 27) puts it, "effectively debellicized" and lead to a paralysis of their foreign affairs. To that effect, a major German weekly paper assessed German society as being "psychologically disarmed" (cited in Kümmel and Leonhard 2005: 524). In the Belgian case, the Rwandan experience provoked the, albeit unofficial, specialization of the military in low-intensity operations, such as humanitarian and peacekeeping missions. According to Manigart (2005: 561), such an increasingly soft power orientation contributed to the Belgium's refusal to deploy troops to Iraq.

Similarly, the German government ruled out the deployment of troops in Iraq while offering to contribute to the reconstruction efforts after the war. It can be assumed that public attitudes played a key role in the political decision-making process. To some extent, the lack of clearly defined criteria about the initiation, continuation or termination of military operations may have led the populace to eschew the involvement of German troops in Iraq. According to Kümmel and Leonhard (2005: 528), such intransparency is detrimental to military legitimacy and the public commitment to out-of-area missions. It may also open into the widespread refusal of an active foreign and security policy. The simultaneous predominance of domestic policy issues in the public mind indirectly compounds the general incomprehension of the risks that German soldiers face in military operations, which appear ostensibly unrelated to the pursuit of national security and prosperity (Kümmel and Leonhard 2005: 529). The elaboration of national interests within the domestic political process influences the legitimacy of military operations and equally the "price the nation is prepared to pay" (Smith 2005: 497). However, when the true national interests are unclear, contradictory to public preferences or "somewhat hidden behind humanitarian arguments" (Kümmel and Leonhard 2005: 523), as in the case of Germany, the deployment of soldiers may be publicly questioned. As to a general conclusion, it becomes evident that governments base their decision-making about military interventions on multiple considerations in the realms of domestic politics and foreign relations. The influence of the casualty factor arises from a complex set of, at times conflicting and not solely demography-related, factors.

3.4 Conclusion

Demographic trends differ markedly between world regions. This global demographic divide may become security-relevant when youth bulges and other imbalances in the population structure exacerbate existing socioeconomic scarcities and resource vulnerabilities in less developed regions, and when demographic aging undermines the capacity of industrialized states to respond to these security risks with adequate development cooperation or military intervention. On a broader scale, this detrimental effect results from changes in three important conditions of foreign policymaking that are subject to demographic trends, namely politics, fiscal resources and public opinion. Even worse, while the demand for Western security capacity shows no signs of abating, their demographic base for military recruitment is in decline. Matching manpower demand and supply will likely become more difficult in the near future.

Chapter 4
The Selectivity of Manpower Demand and Supply

4.1 Human Factors in the Modern Battlefield

The dominating view during the Cold War was that manpower issues played a subordinate role in strategic studies and were seen as "virtually irrelevant" (Freedman 1991: 8) for military power and security policy. Ever since there has been a whole stream of literature on the link between personnel characteristics and military effectiveness (Warner and Asch 2000: 96). In that, force quality is typically assessed by the level of educational attainment and cognitive ability, as measured by individuals' performance on the AFQT (Armed Forces Qualification Test) and in-service skill qualification tests (NRC 2003: 25).

Although enlistment standards may vary over time – either due to a modified military role set or in response to changes in the external recruiting environment – recruits with a high school degree and scores in the upper half of the recruitment cycle's AFQT score distribution are commonly considered as high quality (Hosek 2003: 190). The underlying assumption is that applicants with a high school diploma and above-average entry test scores will perform better as soldiers given their proficiency and trainability (Horne 1987: 443). In fact, however, these entry-level measures serve as proxies for unobservable personal attributes, such as ability, motivation, physical coordination, effort, ability to work in teams, and other job-specific skills that will ultimately determine an applicant's productivity (Asch et al. 2005: 9).

The redefinition of military roles and mission objectives led to a shift in soldierly tasks and an expansion of soldierly requirements. According to Haltiner and Kümmel (2009: 75), the capacity of the military to realize its commissioned functional role set depends, as in any other institution, on the skills and capabilities of the individuals that make up the institution. Thus it appears that the "human factor", i.e. the human capital of each single soldier determines the level of overall military effectiveness and thereby, "the gap between pretensions and reality regarding the realization of the tasks and functions commissioned and entrusted to the institution". Similarly, Biehl

W. Apt, *Germany's New Security Demographics: Military Recruitment in the Era of Population Aging*, Demographic Research Monographs, DOI 10.1007/978-94-007-6964-9_4, © Springer Science+Business Media Dordrecht 2014

et al. (2000: 12) argue that the human factor plays a decisive role in the success or failure of contemporary military operations.

The operational environment of contemporary military operations is characterized by a high degree of complexity, uncertainty and risk, often lacking a clear mandate and well-defined rules of engagements (van Bladel 2004: 12). Missions in such a fraught and "fuzzy" environment (Manigart 2005: 573) require soldiers to switch from high-intensity combat to peacekeeping roles at a moment's note, placing extraordinary demands on the individual and making the case for "high-quality general-purpose forces" (Boot 2005: 108).

In view of the high degree of uncertainty about the nature of evolving security threats and the future use of military force, as well as the continuing sophistication of military technology, Hosek (2003: 181) calls for military personnel "who can learn rapidly, reach high levels of competence, adapt in the face of uncertainty, and apply a variety of skills in difficult circumstances". Hosek (2003: 187f.) further elaborates that versatility, which he defines as the capability to carry out a broad range of tasks, and leadership skills, that means the ability to prioritize objectives, allocate resources efficiently towards them, enhance unit cohesion and motivate personnel, constitute the two key requirements for soldiers in the twenty-first century. Similarly, the U.S. Quadrennial Defense Review Report (U.S. Department of Defense 2001: 9) outlines contemporary soldierly requirements and draws the link between force quality and mission effectiveness:

> The Department of Defense must recruit, train, and retain people with the broad skills and good judgment needed to address the dynamic challenges of the 21st century. Having the right kinds of imaginative, highly motivated military and civilian personnel, at all levels, is the essential prerequisite for achieving success. Advanced technology and new operational concepts cannot be fully exploited unless the Department has highly qualified and motivated enlisted personnel and officers who not only can operate these highly technical systems, but also can lead effectively in the highly complex military environment of the future.

According to Däniker (1992: 185), the principal tasks of the soldier in the twenty-first century are to "protect, help and save". Therefore, it appears that mere warriors, combatants and technicians are of no avail in contemporary military operations, while there is a great demand for a new type of soldier, which Däniker (1992: 185) refers to as "miles protector" and perceives as a soldier, who, while on an assignment abroad, features classical combat strength along with the capacity to protect, assist and respect the local population. Hence, the range of soldierly tasks tends to be complex and diffuse, which manifests in a role set that is no longer limited to genuine military specialties. The resultant job design of modern soldiers may be summarized as follows (Haltiner and Klein 2005: 15): "Nowadays, soldiers primarily act as policemen, guardians, diplomats, judges, referees, paramedics, administrators, and social workers, but hardly ever in the role for which they were trained, namely as soldiers."

The changes in the military role set and nature of military operations have been attended by a sweeping redefinition of soldierly requirements, now accentuating the need for professionalism, social skills, value competence along with combat-related traits like physical fitness and mental resilience. The underlying character traits and technical skills are summarized in Fig. 4.1 and will be discussed in greater detail below.

Professional competence	Social competence
Technical and manual skills	Overall type of personality
Analytical and conceptual skills	Interpersonal skills
Foreign language competence	Coping with conflicts and crises
	De-escalation skills
	Organizational skills
	Talent for improvisation

Value competence	Physical and mental strength
Ethics and morale	Health status
Integrity	Deployability abroad
Intercultural skills	Body fitness
	Stress resistance
	Stamina and firmness
	Coping with psychological strain

Fig. 4.1 Requirements for the modern soldier (Source: Beck 1998: 26–29)

4.1.1 Professional Competence

Professional competence subsumes soldiers' technical know-how, analytical skills and foreign language ability (Beck 1998). The development of these skills takes place in school and their appropriateness for military service is tested during a range of entrance examinations.

The recruitment focus on youth with ample human capital endowments, which has been observable in the U.S. since the early 1980s, mainly stems from considerations of productivity and manpower readiness. It has been argued that "high-quality personnel outperform low-quality personnel" in every key measure of military productivity (Hosek 2003: 192).

For example, first-term military attrition is lower and promotion times are shorter for soldiers with more years of education and higher AFQT scores (Smith et al. 1991: 62f.). Moreover, high-quality personnel appear more productive than low-quality personnel (Warner and Asch 1995: 369). In a study of the operational availability of ships in the U.S. Navy, it has been shown that the material condition of shipboard equipment is a function of soldiers' cognitive ability and training received in the U.S. Navy. Ship downtime was found to decrease as the ratio of high-school graduates aboard increased (Horowitz and Sherman 1980: 68). A subsequent study of ship readiness confirmed the positive impact of personnel quality on every dimension of readiness, more precisely the equipment failure rate, repair rate, and equipment condition (Junor and Oi 1996: 13).

Similarly, it has been shown that the level of aptitudes, as measured by AFQT scores, predict soldiers' performance in teams of signal personnel, more precisely their ability to install and operate communication networks. "Smarter" signal operators are better able to perform team tasks associated with combat missions and provide usable battlefield communication systems. Conversely, lower military enlistment standards of cognitive ability may degrade the team performance of signal operators and hence jeopardize the availability of communications systems providing command and control in combat situations (Winkler 1999: 420f.). Another investigation into the productivity of gunners and tank commanders, who both interact in the handling of weapon systems, showed that on-the-job performance, as measured by tank firing scores, was substantially better among "smarter" tank crew members with higher AFQT scores (Scribner et al. 1986: 201). While military experience has been found to work as a "partial equalizer" for low-ability tank crews, Scribner et al. (1986: 202) caution that experience cannot compensate for low mental ability. Similar tradeoffs between soldier quality and combat performance have been found by Orvis et al. (1992), this time with respect to air battle outcomes. They showed that the level of proficiency among enlisted personnel has a direct and consistent influence on army mission performance, in particular battlefield survival, and other tactical and technical dimensions of air combat (Orvis et al. 1992: 23–47).

The advances in weapon technology have two-fold implications for manpower requirements: While the design is aimed at an improved accessibility of the equipment for non-specialists and a generally simplified mechanical handling, they also create novel complexities in the realm of logistics, doctrine, coordination, intelligence, command and control. As a result, soldierly requirements in terms of educational attainment and on-the-job training will in all likelihood continue to increase (Segal 1986: 16–20). Similarly, Scribner et al. (1986: 202) point out that the need for technical proficiency is particularly pronounced in combat situations when the availability of weapon and support systems may be degraded through heavy duty or battle damage. In such a situation, mission effectiveness depends on high quality logistical support, in the form of base maintenance and immediate repairs under battlefield conditions. According to Handel (1981: 242), this will put higher demands of technical proficiency on manpower for support operations, as well as command and control functions on all levels. In particular, however, it will raise soldierly requirements for the lower-ranking officers and enlisted personnel. It becomes evident that the increased reliance on networked information systems during operations adds to the military's need for skills and expertise in information technology (Williams 2004: 13; Handel 1981: 242). This demand for technically gifted personnel is particularly urgent in view of the fact that the share of technical jobs among enlisted ranks is about twice as large as the share of technical jobs in the civilian economy (Binkin 1986: 8).

4.1.2 Social Competence

There has been an increasing awareness that military effectiveness within the contemporary functional role set relies on the soldiers' endowment with leadership ability, creativity, initiative, courage, oversight, adaptability and mediation skills (Sandler and Hartley 1995: 158; Williams 2004: 13). In order to reinstate peace and restore internal stability in crisis-ridden areas, Western soldiers are mandated to use force and conduct combat operations and fulfill a wide range of civilian tasks, including the reconciliation between conflict factions, protection of the civilian population, provision of humanitarian assistance, management of post-conflict reconstruction activities, and co-operation partner for non-governmental organizations in the course of military interventions (Schreiner 2005: 14).

In view of the humanitarian character of many missions, military leaders caution to keep a fair balance between selecting and training manpower for war and for other activities (Williams and Gilroy 2006: 102). This is particularly urgent in view of the fact that the traditional military rationale of war fighting and deterrence is not outdated but rather complemented by nontraditional tasks that will shift the soldierly job profile towards that of a "armed global street worker", who, due to the inseparability of traditional and nontraditional military roles, needs to "know how to fight, how to establish local security, how to deal with the local adversaries, and how to cooperate with local partners and civilian international relief organizations" (Kümmel 2003: 432). As a consequence, there will be an increased demand for soldiers with interpersonal skills, foreign language abilities, cross-cultural sensitivity, political knowledge, and diplomatic talent (Williams and Gilroy 2006: 102).

4.1.3 Value Competence

The modifications in the military role set and nature of military operations equally increased soldierly requirements for value competence that, according to the taxonomy of Beck (1998), includes a sound ethical value system, integrity and intercultural skills. The increased demand for such traits becomes evident on four military organizational levels: unit, base, multinational force, and operations abroad (Beck 1998: 37). On the unit and base level, there is an increasing demographic, social, cultural, and religious diversity among soldiers that reflects the heterogeneity of the population. Within contemporary multinational military operations, there is a high degree of cultural diversity occurring both on the command and operational level that soldiers have to cope with in order to secure mission success.

For example, peacekeeping operations of the United Nations are commonly conducted by largely inhomogeneous troops from cross-cultural and cross-national backgrounds. Such differences have ramifications for motivation, leadership and organization, potentially jeopardizing operational effectiveness due to cultural clashes along five value dimensions (Hofstede 1984: 81–84): (1) power distance, i.e. the acceptance of an unequal power distribution; (2) uncertainty avoidance, i.e. the discomfort with uncertainty and preference for stability; (3) individualism versus collectivism, i.e. the value of individual rights and welfare versus group membership for the collective good; (4) masculinity versus femininity, i.e. the bias towards either "masculine" values of assertiveness, competiveness, and materialism or towards "feminine" values of nurturing, the quality of life and relationships; and (5) long-term versus short-term orientation, i.e. the value of long-standing, as opposed to short term, traditions and values.[1] Soldiers' willingness to accept culture-related differences in value judgments, customs and approaches will have a decisive impact on troop cohesion and organizational effectiveness.

Likewise, military missions themselves have become increasingly varied and primarily take place in a foreign cultural environment that is often devastated by war and largely unfamiliar to Western soldiers regarding political, historical, religious, and social distinctions. In such a complex setting, there is an imperative for rapid decision-making on issues of significant magnitude, and with the structural reorganization of many Western militaries, this responsibility is shifted to lower levels of the hierarchy (Dandeker 1995: 32). As a result, the personnel requirements for junior officers and enlisted personnel are increasing, and these ranks will be forced to possess skills and exercise judgments that were formerly characteristic for more senior personnel (Cote 2004: 66). Moreover, the attainment of military objectives appears to be less the result of combat strength or sheer technological superiority but of optimal behavior in strategic and ad-hoc decision-making tasks (Biehl et al. 2000: 12). At the same time, singular cases of misconduct may jeopardize the attainment of mission objectives altogether.

Indicative of the downward shift of responsibility in the command structure and the imminence of situations that require ad-hoc, value-oriented decision-making is the incident of a Bundeswehr master sergeant and his battalion of 19 soldiers that were confronted with a furious crowd of about 200 Kosovo-Albanians when protecting the national heritage of fourteenth century Orthodox Archangels Monastery near Prizren in Serbia (Schreiner 2005: 11). Without the possibility of consulting with the operation command about the appropriate course of action, the master sergeant decided to refrain from an offensive confrontation with the mob, took along the resident monks and left the scene with his battalion so that the monastery got burned down (Schreiner 2005: 11). While the sergeant's approach was not without criticism, it illustrates the potentially long-range political or socio-cultural implications of ad-hoc decision-making on lower levels of the command structure.

4.1.4 Physical and Mental Strength

In light of the heavy emphasis on professional and social skills in the soldierly task profile, Rosenau (1994: 47) advocates to shift the recruitment focus away from physical abilities to other competencies, most notably manual and technical skills. However, on closer inspection, it appears that the military profession, contrary to most civilian jobs, entails extreme physical and mental conditions, which primarily arise during the frequent mobilizations and may include sensory overload, sleep deprivation, supply shortfalls, and exposure to difficult terrain or extreme climatic

---

[1] Hofstede's findings on the impact of culture differences on preferences in terms of management style and work environment are well-established in business-related research on managing organizations with cross-national and cross-cultural diversity.

temperatures (Krueger 2001 cited in Sumer 2007: 3J-12). Obviously, these conditions can only be mastered by individuals with a high degree of physical fitness and mental stamina (McLaughlin and Wittert 2009: 3). Thus, albeit the increasing technologization of military affairs, mission readiness continues to be of a function of the health status of each service member and human factors like age, fitness and body composition that are reflected in soldiers' vigor, perseverance, pace, and coordination skills (Rohde et al. 2007: 139).

The military employs a range of fitness and body composition standards in order to select individuals that appear well-suited to the physical demands of military service. In the case of body dimensions, the underlying assumption is that a proper weight-to-height measure is indicative of good health and physical performance capability (McLaughlin and Wittert 2009: 1). Conversely, the prevalence of overweight and obesity has been associated with greater inability to meet military requirements and readiness obligations (McLaughlin and Wittert 2009: 4). Specifically, excessive weight has been found to reduce soldiers' load carrying and lifting ability, postural stability, mobility, flexibility, balance control and fine motor performance (McLaughlin and Wittert 2009: 5). Therefore, increasing prevalence rates of overweight and obesity among young adults could make recruitment and the maintenance of military effectiveness ever more difficult.

In light of the high prevalence rates of smoking among soldiers and the avoidability of related health effects, the U.S. Institute of Medicine extensively reviewed the effects of tobacco use on military readiness and performance in a recent report (IOM 2009). The report's major findings demonstrate that smoking may impair the physical endurance and performance capacity of soldiers by reducing the oxygen-carrying capacity of the blood. Smoking may also hamper the general visual performance and lead to slower dark adaptation and poorer night vision. Moreover, nicotine-deprived soldiers have been found to suffer from acutely impaired vigilance, dexterity and overall cognitive ability. Given the associated increases in reaction time, these withdrawal effects are particularly relevant to aviation performance and motor-vehicle driving. Tobacco use has also been associated with accelerated hearing loss during aging. Lastly, among recent military enlistees, smoking has been found to be a reliable predictor of early discharge and exercise-related injuries for both males and females (IOM 2009: 34). It becomes obvious that, apart from the well-known longer-term adverse health effects of tobacco use, the short-term impact of smoking and particularly nicotine deprivation may hamper soldiers' performance and jeopardize effectiveness during operations.

4.2   A Typology of Military Task Profiles

Despite general agreement on the superior performance of high aptitude personnel, there have been concerns over whether the emphasis on the recruitment of high-quality youth has been worthwhile given the higher costs of accession and their comparatively low retention rates after the end of the first term (Warner and Asch

Provision of spirit and purpose

Progressive-rational	Conservative-traditonal
Analyzing theater of war in rational and analytical terms and answering for democratic norms	Representing the traditional soldierly ethos and institutional legacy

Practical implementation of orders

Technocratic-bureaucratic	Atavistic-destructive
Responding to operational and technical demands in the battlefield	Executing robust mandates on the ground, potentially with direct enemy encounter

Fig. 4.2 Typology of military task profiles (Source: Kutz 1998: 9)

2000: 97). Hence, it appears equally important to secure an appropriate force mix of higher and lower quality soldiers. This matters most for two reasons: cost effectiveness and military cohesion.

In a typology of contemporary military task profiles, Kutz (1998: 9) differentiates between occupational activities in the Bundeswehr that either aim at the provision of spirit and purpose or pursue the practical implementation of orders (see Fig. 4.2). Job descriptions that fall in the former category are either progressive-rational or conservative-traditional, while occupational images that fall in the latter group are either technocratic-bureaucratic or atavistic-destructive (i.e. "primitive"-destructive).

Soldiers in progressive-rational occupations operationalize the military role set and unfolding events on the battleground against the backdrop of international politics and serve as counselors to their governments. Soldiers in conservative-traditional functions are characterized by military virtues like dutifulness, comradeship, bravery, and moral strength. Soldiers in technocratic-bureaucratic domains are commonly assigned to tasks typical of a manager, technician or expert that fills politically specified orders. Lastly, soldiers in atavistic-destructive functions are primarily found in combat units with a specialized task profile, such as paratroopers, undersea divers, or commando troops, which are not intended for extended warlike operations but require superior abilities in specific situations (Kutz 1998: 10f.). These military task profiles are associated with varying cognitive, social, and physical soldierly requirements.

4.3   The Manpower Demand of the Bundeswehr

The labor market of military manpower demand and supply follows the principles of a monopsony. Manpower resources are captive in a market of a single buyer of military manpower and below-competitive wages presuppose a high degree of labor immobility (Cain 1986: 718). This is particularly relevant in occupations with specific skills that are of limited use outside the military (Sandler and Hartley 1995: 160). Recruitment is heavily focused on the national labor market and youth between ages 18 and 24 years. Under these conditions, the labor demand function is determined by output, such as the security threat level, and input factors like relative wages, factor prices and technical opportunities for factor substitution (Ridge and Smith 1991: 291f.).

The size of military manpower demand is correlated with the broad geostrategic and technological environment, and is also a result of national security strategies, foreign policy goals and military manpower management (Williams and Gilroy 2006: 99). The nature of manpower demand differs between service branches, with the army being the most labor intensive division, while the navy and the air force display a high capital intensity (Sandler and Hartley 1995: 160f.). Thus, military manpower demand is not fixed but depends on the mission, internal factors related to labor productivity, technological progress and the structure of the military organization. Military manpower demand can be broken down in quantity and quality requirements.

4.3.1   Quantity

With a medium-term target size of about 185,000 soldiers, consisting of up to 170,000 professional soldiers and up to 15,000 voluntary basic service recruits, the Bundeswehr has an annual personnel demand of 17,000–18,000 soldiers (Kujat 2011: 6). In order to fill these positions with quality personnel, the optimal application-to-hire ratio would be 3:1. Hence, more than 50,000 people would need to apply. Military planners argue that a lower number of applicants is equivalent to cutbacks in recruit quality and soldiers' on-the-job performance. However, said personnel demand only refers to the segment of professional soldiers. When voluntary recruits are concerned, another 10,000 applicants are necessary to adequately fill the open positions (Buse 2011).

4.3.2   Quality

Applicants for a soldier occupation must meet the basic requirements regarding age, German citizenship, educational attainment, and physical ability (summarized in Cohn 2007: 117–124). Entry into the four career tracks of the Bundeswehr is governed by varying threshold criteria with an emphasis on differences in occupational and educational certification. Applicants for the commissioned officer track (*Offiziere*) must have successfully completed upper level education (*Abitur*) and be within the age range of 17–25 years (or 17–32 years for those with an academic degree). Entry into the track of the traditional non-commissioned officers (*Feldwebel*) either requires completion of middle level education (*Realschule*) or lower level

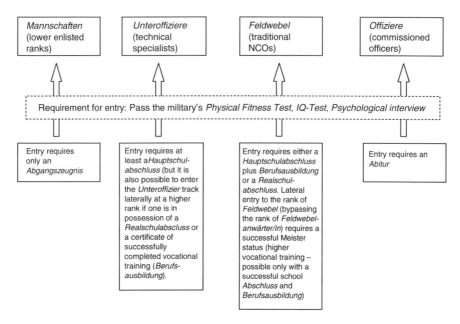

Fig. 4.3 Personnel requirements for entry into the Bundeswehr (Source: Cohn 2007: 117–124)

education (*Hauptschule*) plus a certificate of successfully completed vocational training. In order to become technical specialists (*Unteroffiziere*), applicants must at least have completed compulsory education. If in possession of a lower secondary degree or an occupational certificate, applicants may also laterally enter a higher career rank in the technical specialist track. The mandated age range for applicants interested in the traditional non-commissioned officer and technical specialist career track is 17–25 years (or 17–32 years for those with completed vocational training). Applicants for the lower enlisted ranks (*Mannschaften*) must have completed compulsory education and be between 17 and 32 years old (see Fig. 4.3).

In view of the nature of contemporary military interventions, the Bundeswehr is seeking soldiers with cutting-edge professionalism and motivation backed by sound intellectual abilities, interpersonal skills, physical capabilities, manual and technical skills, and a firm moral concept (Schreiner 2005: 14; Fleckenstein 2000: 87). The diversity of soldierly roles and the extraordinary, rapidly changing circumstances in conflict situations require soldiers to have a sound value system and moral concept based on an appreciation for respect, tolerance, loyalty, justice, cooperation, compromise and empathy that enables soldiers to tell from right and wrong, decide on action or non-action without delay, and act in a generally trustworthy, altruistic and non-partisan way. Moreover, all-purpose virtues like determinedness, perseverance, diligence, fortitude and discipline may help soldiers to endure and complete the mission objectives even under extreme conditions like physical hardship or mental strain (Schreiner 2005: 14).

In addition to these normative qualities, contemporary soldierly standards include mediating and diplomatic skills, negotiating skills, argumentative skills, persuasive power, problem-solving competence, confidence-building ability, as well as other occupational characteristics like foreign language competence, political education and intercultural competence (Schreiner 2005: 14ff.). These soldierly requirements altogether appear in stark contrast to the traditional soldierly requirements of unquestioned obedience and strict compliance with orders of the military command. Similarly, Dalgaard-Nielsen (2006: 109) comments the preferred attributes of Bundeswehr soldiers, including "creative, politically insightful, diplomatic, open minded and culturally sensitive" with the remark: "These may not be the first adjectives springing to the mind of an external observer when thinking about soldiering."

4.4   The Selectivity of Manpower Supply to the Bundeswehr

The long-standing argument in favor of military conscription was that it would ensure an equal representation of the general population within the military, thereby fairly sharing the risks and costs of war among all parts of society (Smith 2005: 503) and working as a "social leveler" (Segal and Segal 2004: 24). By contrast, the all-volunteer force was suspected to reinforce social distinctions by recruiting an unrepresentative force and distributing the burden of war unevenly among certain groups of society (Smith 2005: 503).

Research on young enlistees in the U.S. Army prior to the end of conscription demonstrated their overall representativeness of the nation in terms of educational attainment, intellectual ability, and family background (Mare and Winship 1984: 41). Later U.S.-based research subsequent to the inception of the all-volunteer force proved that the military had become less representative and that men who voluntarily enlisted in the military were a select subset of the male population, with military enlistment being concentrated among those of less privileged socio-economic backgrounds and lower qualifications (Teachman et al. 1993: 287).

The high number of international military engagements and the rising death toll among Western forces have equally raised criticism that it is mainly the uneducated, deprived share of the population that serves in the military and therefore bears the greatest burden of war (Korb and Duggan 2007: 467). However, recent analyses of the demographic characteristics of U.S. military recruits have largely proved otherwise. For example, Segal and Segal (2004: 3) summarized their research on the U.S. military population by stating: "The all-volunteer military is more educated, more married, more female, and less white than the draft-era military." Similarly, Kane (2006) compared the demographic composition of enlisted recruits with the general U.S. youth population in terms of household income, educational attainment, race/ ethnicity, and regional/rural origin. He found that military recruits were, on average, "more highly educated than the equivalent population, more rural and less urban in origin, and of similar income status (Kane 2006: 2). Based on the demographic evidence, Kane (2006: 15) arrived at the conclusion that the U.S. military is "more similar than dissimilar to the general population."

In contrast to the United States, Germany continued to rely on conscription until 2011. However, the number of approved conscientious objectors was high for many years, the conscript ratio (i.e. the ratio of conscripts to regular soldiers within a military organization) amounted to just about 36 % at last (IISS 2008: 124), and the majority of personnel were regular soldiers. On this account, the Bundeswehr was denoted a "pseudo-conscript force" (Haltiner 1998: 16) and would have also satisfied the notion of a "quasi-professional" force.

Despite the selectivity of conscription and the performance-oriented entry requirements, Wolffsohn (2009) recently claimed in a major German newspaper that the Bundeswehr is becoming a "lower class military", which he traced back to the fact that about 35 % of the soldiers would originate from East Germany, leaving them overrepresented in the Bundeswehr with respect to their share of about 20 % in the general population. While Wolffsohn (2009) may be correct to point out this demographic imbalance in view of legitimate equity concerns, his assertions about the association of social origin and human capital endowments of East German soldiers are not consistent with empirical fact.

In the following, based on recent and representative data from the German Microcensus, the socio-demographic composition of enlisted military personnel (i.e. professional or regular soldiers or basic military conscripts) will be compared with that of the general population that pursues civilian employment (i.e. civilian employment or alternative civilian community service) considering age, local origin, educational attainment, and health status. In order to make inferences about the selectivity of military service among youth, the main focus of the analysis rests upon the primary target age group of the Bundeswehr, that is people between 19 and 22 years.[2]

4.4.1 Civilian Community Servants vs. Basic Military Conscripts

Figure 4.4 shows the age composition of young males undergoing civilian community service and basic military service between 1998 and 2005. In both services, the age structure is fairly similar with the majority of males rendering their service to society between the ages 19–22 years.

Table 4.1 reports the share of males from East and West Germany that undergo civilian community service or basic military service. In order to get an idea of the representativeness, it is worthwhile to consult the distribution of East and West Germans in the general population that are liable for both services, namely males with German citizenship. Therein, according to the Microcensus, the share of East

---

[2] The reader is referred to Chap. 5 ("Determinants of Military Manpower Supply") for more information about data, measures and sample composition.

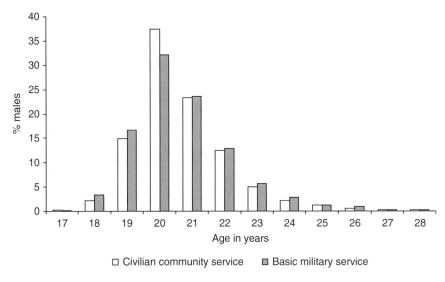

Fig. 4.4 Age structure, civilian vs. military service (Source: Scientific use files of the German Microcensus 1998–2005, own estimations)

Table 4.1 Share East vs. West German origin, civilian vs. military service

Origin	Civilian community servants	Basic military conscripts
East Germany	22.9	38.1
West Germany	77.1	61.9

Source: Scientific use files of the German Microcensus 1998–2005, own estimations

Table 4.2 Share rural vs. urban origin, civilian vs. military service

Settlement size	Civilian community servants	Basic military conscripts
>20,000	56.6	49.5
<20,000	43.3	50.6

Source: Scientific use files of the German Microcensus 1998–2005, own estimations

German males amounts to 28.5 % and the share of West Germans to 71.5 %. Turning to the respective distributions in Table 4.1, it becomes evident that East German males are underrepresented among civilian community servants and overrepresented among basic military conscripts.

Table 4.2 informs about the distribution of rural and urban residents among civilian community servants and basic military service conscripts. When considering the population liable for both services, as recorded in the Microcensus, the share of males from presumably rural settlements with less than 20,000 inhabitants ranges around 47 %. It appears, therefore, that males from rural regions are underrepresented among civilian community servants and overrepresented among basic military service conscripts.

Table 4.3 Educational attainment, civilian vs. military service

Educational attainment	Civilian community servants	Basic military conscripts
Lower level (Hauptschule)	14.6	23.8
Middle level (Realschule)	31.8	41.6
Upper level (Gymnasium)	53.6	34.6

Source: Scientific use files of the German Microcensus 1998–2005, own estimations

Table 4.4 Educational attainment, basic military conscripts, East vs. West

Educational attainment	East Germany	West Germany
Lower level (Hauptschule)	11.6	31.2
Middle level (Realschule)	53.0	34.7
Upper level (Gymnasium)	35.5	34.1

Source: Scientific use files of the German Microcensus 1998–2005, own estimations

Table 4.3 provides an account of the distribution of educational attainment among civilian community servants and basic military service conscripts. Quite noteworthy is the larger share of higher educated civil community servants and the large share of males with middle level education undergoing basic military service. With reference to Wolffsohn (2009), Table 4.4 differentiates military conscripts by East and West German origin, and educational level. In contrast to the claimed relative inferiority of military personnel from East Germany, empirical evidence demonstrates a more favorable distribution of educational attainment among East German conscripts in terms of a higher share of males with middle or upper level education and a relatively small portion with low education.

With regard to the physical performance potential, Fig. 4.5 reports the body composition of males undergoing civilian community service and basic military service. It appears that civil community servants have a somewhat better physique given the higher share of males with normal weight. In the meantime, Fig. 4.6 reports the respective smoking behavior of civil and military servants and shows that the relative distribution of non-smokers is about the same, albeit a higher level of ex-smokers among civilian community servants.

4.4.2   Civilian vs. Military Labor Force

The population of regular and professional soldiers active in the Bundeswehr is markedly younger than the civilian labor force. As shown in Fig. 4.7 the military workforce is mainly concentrated in the younger age groups. With youth entering the services typically shortly after high school, the comparatively smaller shares of military personnel beyond the ages 25 and 32 years can be traced back to the Bundeswehr's personnel policy of employing only relatively few soldiers beyond the status of an extended-service conscript that has a retention period of 4 years or

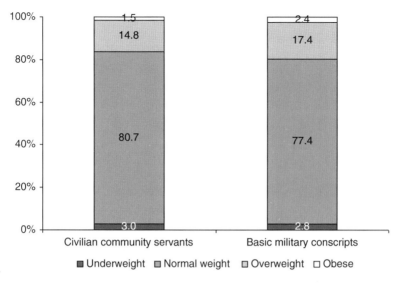

Fig. 4.5 Body composition, civilian vs. military service (Source: Scientific use files of the German Microcensus 1999, 2003, 2005, own estimations)

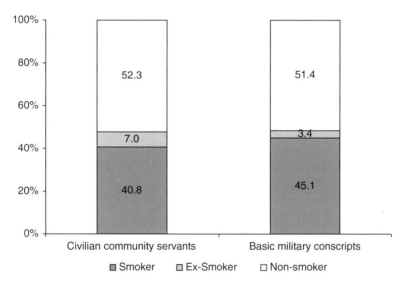

Fig. 4.6 Smoking behavior, civilian vs. military service (Source: Scientific use files of the German Microcensus 1999, 2003, 2005, own estimations)

beyond the temporary-career volunteer status with a retention period of 12 or 13 years. By contrast, the age structure in the civilian labor force is well-balanced and includes a large share of people that are still economically active beyond the age of 55.

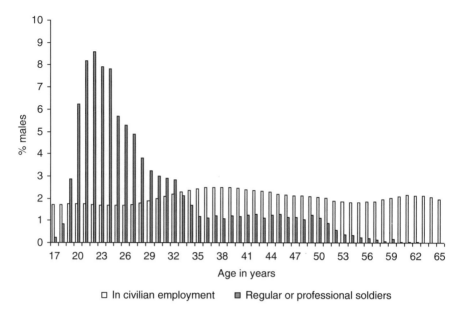

Fig. 4.7 Age structure, civilian vs. military labor force (Source: Scientific use files of the German Microcensus 1998–2005, own estimations)

Table 4.5 Share males vs. females, civilian vs. military labor force

Sex	In civilian employment	Regular or professional soldiers
Female	49.4	4.9
Male	50.6	95.2

Source: Scientific use files of the German Microcensus 1998–2005, own estimations

Table 4.6 Share German vs. foreign population, civilian vs. military labor force

Nationality	In civilian employment	Regular or professional soldiers
German	89.7	100.0
Foreign	10.3	0.0

Source: Scientific use files of the German Microcensus 1998–2005, own estimations

Table 4.5 reveals the stark contrast in sex-specific participation rates in the civilian and military labor force. Accordingly, about half of the civilian labor force at ages 19–22 years is female, while the respective share in the military is just about 5 %. A similar demographic imbalance concerns the portion of foreign nationals in the civilian and military labor force that is represented in Table 4.6. While foreign workers make up about 10 % of the civilian labor force at 19–22 years, the share of non-nationals is zero in the Bundeswehr given the entry requirement of German citizenship.

Table 4.7 reports the share of East and West Germans in the civilian and military labor force. In comparison with the shares of East and West Germans in the general population of 26.9–73.1 %, East Germans are slightly underrepresented in civilian

Table 4.7 Share East vs. West German origin, civilian vs. military labor force

Origin	In civilian employment	Regular or professional soldiers
East Germany	25.0	35.6
West Germany	75.0	64.4

Source: Scientific use files of the German Microcensus 1998–2005, own estimations

Table 4.8 Share rural vs. urban origin, civilian vs. military labor force

Settlement size	In civilian employment	Regular or professional soldiers
>20,000	58.4	46.8
<20,000	41.6	53.3

Source: Scientific use files of the German Microcensus 1998–2005, own estimations

Table 4.9 Educational attainment, civilian vs. military labor force

Educational attainment	In civilian employment	Regular or professional soldiers
Lower level (Hauptschule)	26.3	23.9
Middle level (Realschule)	38.8	50.3
Upper level (Gymnasium)	34.9	25.8

Source: Scientific use files of the German Microcensus 1998–2005, own estimations

employment and markedly overrepresented in the military workforce. Turning to the share of persons originating from a rural or urban setting displayed in Table 4.8, it appears that persons from presumably rural settlements with less than 20,000 inhabitants are overrepresented in the military workforce given the respective lower share of 44.4 % in the general population, as reported in the Microcensus.

Table 4.9 provides insight in the distribution of educational attainment in the civilian and military labor force. As previously observed for basic military conscripts, the large portion of military personnel with middle level education is striking. Accordingly, about half possesses middle level education, while the other half splits evenly in military personnel with low or high educational attainment. In the civilian labor force, these shares are somewhat more balanced.

Table 4.10 refers back to the claim made by Wolffsohn (2009) regarding the socio-demographic attributes of the East German military population. As previously outlined for the East German share of basic military conscripts, it appears that East German soldiers display a more favorable distribution of educational attainment than their West German peers. This is again reflected by a markedly higher share of males with either middle or upper level education, as well as the much lower share of East German soldiers with lower level education.

With regard to health-related factors, Fig. 4.8 informs about the body composition of persons in civilian and military employment. It reveals a noticeable difference in the prevalence of normal weight, meaning that a larger portion of the military labor force is of normal weight. This unequal distribution in body composition mainly stems from the high share of underweight persons in civilian employment at 19–22 years. On the other hand, as shown in Fig. 4.9, the share of smokers is markedly higher among the military workforce.

Table 4.10 Educational attainment, regular soldiers, East vs. West

Educational attainment	East Germany	West Germany
Lower level (Hauptschule)	10.4	31.4
Middle level (Realschule)	61.6	44.0
Upper level (Gymnasium)	28.0	24.6

Source: Scientific use files of the German Microcensus 1998–2005, own estimations

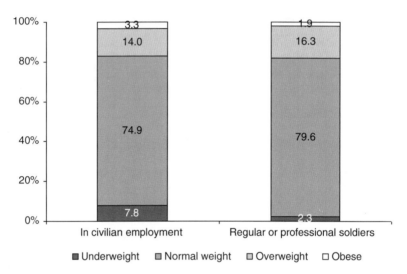

Fig. 4.8 Body composition, civilian vs. military labor force (Source: Scientific use files of the German Microcensus 1999, 2003, 2005, own estimations)

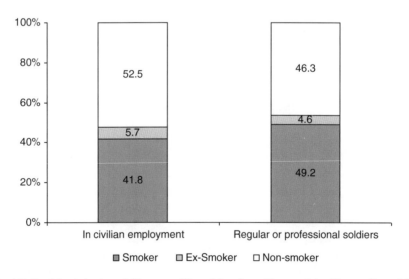

Fig. 4.9 Smoking behavior, civilian vs. military labor force (Source: Scientific use files of the German Microcensus 1999, 2003, 2005, own estimations)

4.5 Conclusion

Taken together, the preceding analysis leads to two broad conclusions: Firstly, there is a distinct selectivity of military service in Germany, regardless of whether the group of basic military service conscripts or professional soldiers is concerned. In line with Kane (2006) or Segal and Segal (2004), the established findings could be summarized by the statement: when military personnel between 19 and 22 years is concerned, the Bundeswehr is more East German, more rural and more educated than the general population, while differences in health-related factors remain inconsistent. Secondly, while the Bundeswehr recruits a disproportionate share of military personnel from East Germany, it enlisted East German males (and females) with comparatively high human capital endowments. This phenomenon of "positive selectivity" into the military is consistent with previous research and has also been observed among the Black male population in the United States (Teachman et al. 1993: 305).

Chapter 5
Determinants of Military Manpower Supply

5.1 Theoretical Background

Factors that influence military recruiting fall into two broad categories (Bicksler and Nolan 2006: 2). Those in the first category are largely outside the military's control, yet have a marked influence on the number and types of youth available for military recruitment. They include the size and composition of the youth population, youth values and their attitudes towards the military, opportunities and options in the civilian labor market or the educational system, as well as trends in student achievements and physical abilities (Boëne 2003: 170). In order to cope with these economic and demographic conditions in the external recruiting environment, the military can employ a range of policy measures that foster manpower supply. These measures, which belong to the second category, include the variation of the number of recruiters, the amount spent on advertising, enlistment bonuses, scholarship programs and military pay (Bicksler and Nolan 2006: 2).

The determinants of military propensity and enlistment have been extensively analyzed over the past decades. This research mainly took place in the United States, where the adoption of the all-volunteer force in the early 1970s echoed in a large body of literature (NRC 2003: 9). According to Berner and Daula (1993: 1f.), most research into the determinants of enlistment supply may be classified into the following major categories: (1) Analyses of time-series or aggregate panel data that provide regression estimates about the influence of various individual and macro level characteristics on the likelihood of military enlistment (e.g. Hosek and Peterson 1985; Asch et al. 1999); and (2) models that focus on the institutional recruiting environment, and in particular, the efficacy of enlistment incentives and recruiter effort (e.g. Dertouzos 1985).

Past research has been highly consistent in identifying variables predictive of self-reported propensity to enlist or actual military enlistment. Macro variables typically include civilian labor market conditions (e.g. unemployment rates), civilian pay opportunities, relative military pay and benefits, and the costs of further education (Warner and Asch 1995: 352). On the micro level, military propensity

W. Apt, *Germany's New Security Demographics: Military Recruitment in the Era of Population Aging*, Demographic Research Monographs, DOI 10.1007/978-94-007-6964-9_5, © Springer Science+Business Media Dordrecht 2014

and enlistment decision-making have been linked to factors like age and race, educational attainment, family background (e.g. household income, number of siblings, parental education), regional provenance, and settlement size (Murray and McDonald 1996: 9). Similarly, individual attitudes toward the military and its current missions, as well as preferred job characteristics influence the attractiveness of military service (summarized in Bachman et al. 2000).

Models of individual enlistment decision-making commonly draw on standard occupational choice theory, originally formulated by Rosen (1986: 644), which differentiates between two sectors of the economy: The military sector and the civilian sector (summarized in Warner and Asch 1995: 352). With reference to the expectancy theory developed by Vroom (1964), individual decision-making related to military service is based on the interplay of expectations and values. In a situation of imperfect information, youth will evaluate the perceived set of rewards and costs associated with military service and other occupational alternatives. In this connection, parents (in particular, fathers that commonly serve as "role models"), friends and other peers have been identified as critical reference points in the occupational decision-making of youth. Hence, Fishbein and Ajzen (1975) argue that there are two primary sources of individual intentions: Personal attitudes and the perception of other people's attitudes, in particular those of people that are close (cited in Nieva et al. 1997: 4).

Military service is associated with some distinct occupational features. Specific requirements that originate from the military lifestyle include the subjection to military discipline, the inability to modify the fixed terms of enlistment, strike or negotiate over working conditions, extended service availability, absences from home, geographic mobility, frequent moves to relatively remote areas, residence in foreign countries, and most importantly the risk of death and injury during military maneuvers and combat operations (Segal 1986: 15). While all of these demands are strenuous for the service member, they also entail a high degree of stress, hardship and disruption for the family. This strain may arise from different sources, including the potential for casualty of the service member, the infringement of spouses' employment opportunities and career continuity, the disruption in the cognitive and social development of children in school, and the spouses' sole responsibility for parenting during times of separation (Segal 1986: 16; Moskos 1977: 42).

On the other hand, the military also provides its personnel with a range of transferable skills that are of significant value in the civilian sector such as driving, computer literacy, technical expertise or pilot training (Sandler and Hartley 1995: 166).[1] The prospect of acquiring these transferable skills and gaining hands-on experience makes military service an attractive career entry (Becker 1962: 16). In addition, military service offers some non-pecuniary rewards over civilian occupations, which include pride of service, opportunity for travel and stable employment (Sandler and Hartley 1995: 156). These non-pecuniary factors have been identified as important enlistment drivers in combination with the broader social, cultural and

---

[1] Other forms of training (e.g. missile operation, shooting or parachuting) are military-specific and less efficient in attracting enlistments.

political circumstances, as well as individual-level motivations like patriotism, family tradition, desire for adventure, and strive for personal growth in light of a meaningful experience (Burk and Faris 1982: 11; Wilson et al. 1988: 2; Battistelli 1997: 471). Moreover, Eighmey (2006) demonstrated that certain international or domestic events, such as wars or terroristic attacks, are additional factors associated with the desirability of a military career.

Individual learning ability (Hosek and Peterson 1985: 9) and aspirations regarding skill enhancement constitute further elements in the career choice process. Youth with above average human capital endowments generally have higher potential returns from further education and better opportunities in the civilian market (Warner and Asch 1995: 354). This also applies to the state of physical and mental health, which Becker (1975: 7–36) and Grossman (1972: xv) defined as additional dimensions of human capital.

Military service will be chosen when (1) its expected utility from pecuniary and non-pecuniary benefits is perceived to exceed that of any other options, (2) the incentive value of human capital development is high, and (3) the applicant believes to possess the abilities to perform the military-specific training (Warner and Asch 1995: 352). Albeit those individual-level determinants of military propensity, becoming a soldier is still a "two-party decision" (Bachman et al. 2000: 4). In matching jobs to people, the military considers various enlistment criteria in terms of educational attainment, cognitive skill, physical ability, and moral standards of the applicant that determine actual manpower supply.

Figure 6.8 provides an abstract overview of the enlistment process and summarizes the main underlying factors. The process is conceptualized in two stages, namely initial propensity formation, i.e. establishment of potential supply based on the likelihood of entry, and the subsequent conversion of that potential supply into military enlistments (Orvis et al. 1996: 2). This conversion process is accompanied by a comparison of applicants' human resources and the specific entry requirements of the military (Bachman et al. 2000: 2) (Fig. 5.1).

5.2 Hypotheses

Building on propositions from the literature, it is possible to generate and test hypotheses about the expected statistical relationship between military enlistment and several underlying factors in the decision-making process.

The selected observation period falls together with the complete opening of all careers and career groups to women in the Bundeswehr in the year 2001. Despite this expansion of women's roles in the Bundeswehr, military service has largely remained a male endeavor. In part, this may also stem from the traditionally masculine culture and structure of the military organization (Segal and Segal 2004: 23).

Hypothesis 1:   *The chances of females to serve in the military are miniscule in comparison with males.*

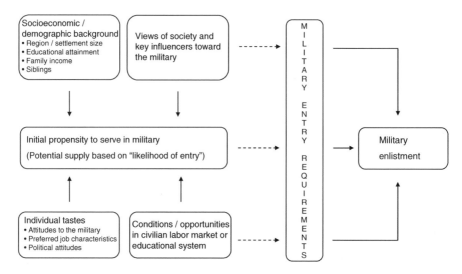

Fig. 5.1 A conceptual overview of the enlistment decision (Source: Adapted from Orvis et al. 1996: 4 and Bachmann et al. 2000: 2)

Various studies have demonstrated the impact of prevailing conditions in the civilian labor market on the likelihood of youth to serve in the military (e.g. Dale and Gilroy 1983; Kleykamp 2006; Kilburn and Klerman 2000). It appears that unemployment in the civilian labor market is positively correlated with military service. At the same time, youth from rural areas have been found to serve in the military at higher rates than their peers from central cities and surrounding suburban areas (O'Hare and Bishop 2006: 1). With the declining employment base in many rural areas and limited resource availability in rural families, military service is mostly a means to an end, either to get some higher education, acquire job-related skills, obtain job security, or get some experience outside the countryside (Carr and Kefalas 2009: 89). Hence, in regions with high unemployment, where civilian jobs are harder to find, youth are more likely to consider military service as an occupational choice, and it is easier to recruit high-quality youth (Bicksler and Nolan 2006: 3). Hence, local origin may also be considered as a predictor of a constellation of other factors that influence the attractiveness of military service, in particular, the local employment outlook for youth and prevailing wage levels.

Hypothesis 2a: *Youth from East Germany are more likely to consider a military career and serve in the military.*

Hypothesis 2b: *Youth from economically weak federal states with high rates of youth unemployment are more likely to consider a military career and serve in the military.*

Hypothesis 2c: *Youth from urban areas, i.e. settlements with a size above 20,000 residents are less likely to consider a military career and serve in the military.*

Youth with more human capital endowment have higher potential returns from further education and generally better opportunities in the civilian market (Sandler and Hartley 1995: 166; Hosek and Peterson 1985: 12). In the transition from school to career, major civilian institutions have been the preferred establishment for the vast majority of youth with an above-average learning proficiency (Sandell 2006: 85). Hence, there appears to exist a negative relationship between military enlistment and individual human capital endowments, in particular academic ability (Warner and Asch 1995: 354).[2]

Hypothesis 3:   *Youth that are better educated are less likely to consider a military career and serve in the military.*

At younger age, when graduation from school is still in the distant future, youth tend to have vague plans in terms of their educational attainment and occupational preferences. As they become older and move towards the end of school, they acquire more information about further education and occupational options. In this process, youth rule out some alternatives and form preferences about their future career (Lawrence and Legree 1996: 2; Bachman et al. 2000: 27). Thus, there appears to be a negative relationship between age and military propensity. The military pays a premium on youth, and the age structure of military personnel is markedly younger than that of the civilian labor force (Segal and Segal 2004: 23). In combination with the lack of lateral entry, the likelihood of entering or serving in the military should decrease with age. However, given the narrow age range of 19–22 years that is under observation and the fact that most soldiers enter the military a few years after graduation from school, there may be a *positive* relationship between age and military service.

Hypothesis 4a: *Individuals at 18–20 years are less likely to consider a military career relative to the younger age group.*

Hypothesis 4b: *Individuals in the most advanced age group of 22 years are most likely to serve in the military.*

Military service imposes great demands on service members' commitment, time and energy. The risk of injury or death, regular separations from the family, frequent geographical relocations and long, often unpredictable, duty hours contradict service members' roles outside the military and especially increase the stress

---

[2] The use of educational attainment as a proxy for learning proficiency appears consistent with the Bundeswehr's use of it as the major entry criterion and a predictor of future performance on-the-job.

and hardship of their families (Segal 1986: 15; see also Segal and Segal 2004: 32 f.).

Hypothesis 5:   *Married individuals are less likely to serve in the military than their unmarried peers.*

Youth that originate from families with lower per capita household income are less able to self-finance higher education (Kilburn and Klerman 2000: 11; Hosek and Peterson 1985: 10). Research has also shown that parents with limited available resources only have "leveled aspirations" for their children and often encourage their offspring to choose full-time employment with a high level of material job security like entering the military (Carr and Kefalas 2009: 60). The issue of resource availability for education is ever more important in families with several children. In line with the human capital theory established by Becker (1975), research has shown that individuals with siblings are more likely to serve in the military than those without (Kilburn and Klerman 2000: 11; Rose 2007: 30).

Hypothesis 6a: *Youth with a per capita household income of less than 300 Euros are more likely to serve in the military than others of higher social origin.*

Hypothesis 6b: *Youth with one or more siblings are more likely to serve in the military than youth that are the only child in the family.*

Contemporary military operations entail high physical demands and frequent deployments. The purpose of military entrance examinations is to select individuals that are characterized by physical strength and endurance (McLaughlin and Wittert 2009: 1). On the assumption that normal weight is an indicator of the overall health status and physical readiness, the Bundeswehr does not accept applicants with a BMI higher than 30 (Leyk et al. 2006: 643). At the same time, as in other Western militaries, the Bundeswehr does not apply an enlistment standard with respect to the use of tobacco or cigarette smoking (cf. NRC 2006: 163). This may stem from the fact that most adverse health effects of tobacco use on health and physical performance tend to emerge only much later in life (NRC 2006: 167). Therefore, the Bundeswehr offers employment to a selective set of males with a favorable body composition, while it does not differentiate applicants by smoking status.

Hypothesis 7a: *Individuals that are overweight or obese are less likely to serve in the military relative to individuals of normal weight.*

Hypothesis 7b: *There is no predictive relationship between smoking behavior and military service.*

During the period of observation, the Bundeswehr underwent a significant institutional transformation that entailed large-scale force reductions. The 1998 personnel strength of 336,243 service members went down to a level of 259,389 in 1998. In this process, the number of regular and professional soldiers decreased from

191,332 in 1998 to 189,431 in 2005. The number of basic military service conscripts decreased from 135,802 in 1998 to 69,958 in 2005 (Federal Ministry of Defense 2009).

Hypothesis 8:   *Time is negatively is associated with the likelihood of military service, in particular, among youth eligible for basic military service.*

5.3 Methodological Approach

To test the hypotheses regarding socio-demographic predictors of military service while controlling for other variables, a broad range of multivariate analyses was performed. The underlying motivation for these was the awareness that conclusions based on descriptive statistics, as presented in the context of the selectivity of manpower supply, may be misleading since the relationships may be spurious. For example, the effect of regional origin and/or educational attainment may be overstated in cross-tabulations because other factors simultaneously influence the outcome variable. It appears that there is not only a relationship between the independent variables and each of the dependent variables, but also among the independent variables. For example, youth enlisted in the military as a soldier or basic military conscript are likely to differ demographically, socio-economically and physically from those that pursue civilian employment or choose community service instead. In order to account for the incomparability among risk factor groups, a series of multivariate logistic regression models was estimated to describe the relationship among a set of background variables and obtain a correct, unconfounded estimate of their combined net effects on the outcome variable (Hosmer and Lemeshow 2000: 73).

With the conceptualization of the outcome variable as a binary response in terms of the generic statistical terms "success" and "failure", a logistic regression model was employed that fits the data on the basis of maximum likelihood estimation (MLE). Applying a logistic regression model allows for an assessment of whether and to what degree a set of independent variables has a predictive relationship to a dichotomous dependent variable (Agresti and Finlay 1999: 577f).[3] Using logistic regressions, the relationship between military propensity and actual military service and a set of independent variables is examined, including individual-level characteristics and family background factors. The outcome variable military propensity is differentiated as follows:

---

[3] Although statistical models like the logistic regression depict correlation among variables, it does not allow inferences about whether a relationship is causal. In consideration of the statisticians' maxim "association does not imply causation", a statistical association between the dependent and independent variables, even if it is significant and/or very strong, does not constitute conclusive evidence that the estimated change of an independent variable actually causes a change in the outcome variable.

1. being interested in the Bundeswehr as a potential employer ("success") vs. not being interested in the Bundeswehr as a potential employer ("failure")
2. having the intention to enlist as a basic military service conscript in the Bundeswehr ("success") vs. not having the intention to enlist as a basic military service conscript in the Bundeswehr ("failure")
3. having the intention to enlist as a regular or professional soldier in the Bundeswehr ("success") vs. not having the intention to enlist as a regular or professional soldier in the Bundeswehr ("failure")

The outcome variable military service is classified by:

4. being enlisted in the military as a regular or professional soldier ("success") vs. pursuing other civilian employment ("failure"), and
5. undergoing basic military training ("success") vs. performing alternative civilian community service ("failure").[4]

5.4   Data and Sample Composition

5.4.1   Military Propensity

Data for analyzing the determinants of military propensity are drawn from the 2006 nationally representative youth opinion poll about the security political situation in Germany. The survey population includes youth between 16 and 20 years that live in private households and count as German residents (AIK 2006). The collection of the survey data followed three steps and was based on computer-assisted telephone interviewing: (1) Based on official register data, data collectors drew a stratified sample of phone numbers representative of the general population in Germany. (2) The final digits of the phone numbers in the preliminary sample were randomized, so that also phone numbers belonging to individuals previously unrepresented in the register data were dialed. (3) The preliminary sample was reconciled with the target population of youth between 16 and 20 years. In the event of multiple individuals fulfilling the sample selection criteria, the interviewee was chosen randomly (AIK 2006). Table 5.1 provides an overview of the socio-demographic composition of the subsamples included in the analysis.

---

[4]The success-failure coding of basic military service and alternative civilian service rests on the assumption that both are mandatory for males and do not constitute an occupational choice as such. Therefore, other types of employment are disregarded in the respective regression models, just like the possibility of being exempted from military (and civilian) service due to a failure of the military medical entrance exams.

Table 5.1 Enlistment propensity, composition of sample

Demographic characteristic	Question: "Are you interested in the Bundeswehr as a potential employer?"		Question: "Do you intend to enlist as a basic military service conscript in the Bundeswehr?"		Question: "Do you intend to enlist as a regular or professional soldier in the Bundeswehr?"	
	No. of observations	% sample	No. of observations	% sample	No. of observations	% sample
Dependent variable						
No	1,093	62.6	313	49.3	173	60.7
Yes	652	37.4	322	50.7	112	39.3
Independent variables						
Sex						
Male	870	49.9	635	100.0	285	100.0
Female	875	50.1	0	0.0	0	0.0
Germany						
West Germany	1,339	76.7	478	75.3	204	71.6
East Germany	406	23.3	157	24.7	81	28.4
Settlement size						
<20,000 residents	758	43.4	263	41.4	120	42.1
>20,000 residents	987	56.6	372	58.6	165	57.9
Educational attainment						
Middle level or below	844	48.4	302	47.6	158	55.4
Upper level	901	51.6	333	52.4	127	44.6
Age group						
16–18 years	1,242	71.2	495	78.0	229	80.4
19–20 years	503	28.8	140	22.1	56	19.7
Total	1,745		635		285	

Source: Youth survey on the security political situation in Germany (AIK 2006), own estimations

5.4.2   Military Service

Present analyses about the determinants of military service are based on the German Microcensus. This survey provides a large number of observations and constitutes one of the few publicly available data sources that accounts for the share of the population in soldierly occupations. The Microcensus is an annual cross-sectional survey of 1 % of German households and has a response rate of about 97 % due to compulsory participation. It provides representative information about the demographic, social and economic circumstances, as well as the education and vocational background of the population (e.g. Shahla et al. 2005). In addition to these major inquiry topics, there is a regularly conducted additional survey program on the health situation of the population.

All analyses draw upon the Scientific Use Files of the Microcensus, which contain a 70 % random sample of the original 1 % sample, as available from the Federal Statistical Office. There is a disproportionate sampling fraction of 0.45 % for variables included in the Labor Force Survey of the European Union. This includes information on respondents' occupational activity specifying military membership as a professional soldier or basic military conscript. Data for the years 1998–2005 are pooled to ensure a large enough sample size and a sufficient representation of the military population. The surveys of 1999, 2003 and 2005 contain a voluntary 0.45 % sample of respondents that provide information about their health.

Analyses are restricted to the German youth population of 19–22 years at the time of the interview in order to capture a large majority of those individuals that have entered military service shortly before and thereby increase the chances that the respondent provides information about his/her family of origin and less so about his/her independently owned living circumstances that are of minor importance in this analysis. Restricting the data to youth with non-missing information on both the dependent and all independent variables of interest leaves a total of 118,932 males and females for analyzing the probability to serve as a regular or professional soldier and a total of 6,405 males for analyzing the probability to serve as a basic military service conscript. In addition, youth that do not live in a private household, are still in school or non-employed are excluded from the sample. Tables 5.2, 5.3, 5.4, and 5.5 provide an overview of the socio-demographic composition of the sample populations under review.

5.4.3 Measures

All regression models that predict military propensity or military service include the independent variables sex, local characteristics in terms of East or West Germany and settlement size, as well as educational attainment and age. In addition, regression models that predict actual military service include marital status, family background characteristics, such as per capita household income and the presence of younger children in the household, and, data availability permitting, respondents' health status as measured by their body mass index and smoking behavior.

5.4.3.1 Military Propensity

Military propensity is measured by several questions asking about the viability of a military career. In a series of items relating to general aspects of their security political interest, youth were asked: "Are you interested in the Bundeswehr as a potential

Table 5.2 Composition of sample, soldiers vs. civilian employment

Demographic characteristic	No. of observations	% sample
Dependent variable		
Soldier	2,037	1.7
Other	116,895	98.3
Independent variables		
Sex		
Male	60,464	50.8
Female	58,468	49.2
Germany		
West Germany	85,092	71.6
East Germany	33,840	28.5
Federal state		
Baden-Württemberg	13,204	11.1
Bavaria	17,918	15.1
Berlin	5,201	4.4
Brandenburg	4,646	3.9
Bremen	1,353	1.1
Hamburg	4,351	3.7
Hesse	6,625	5.6
Mecklenburg	4,669	3.9
Lower Saxony	9,862	8.3
North Rhine-Westph.	17,282	14.5
Rhineland-Pfalz	6,090	5.1
Saarland	3,514	3.0
Saxony	9,142	7.7
Saxony-Anhalt	5,060	4.3
Schleswig-Holstein	4,893	4.1
Thuringia	5,122	4.3
Settlement size		
<20,000 residents	53,115	44.7
>20,000 residents	65,817	55.3
Educational attainment		
Middle level or below	81,371	68.4
Upper level	37,561	31.6
Age group		
19 years	23,874	20.1
20 years	30,313	25.5
21 years	32,333	27.2
22 years	32,412	27.3
Marital status		
Unmarried	114,891	96.6
Married	4,041	3.4
Per capita household income		
<300 Euros	21,252	17.9
>300 Euros	97,680	82.1

(continued)

Table 5.2 (continued)

Demographic characteristic	No. of observations	% sample
Children in the household		
No child <18 years	88,351	74.3
1+children <18 years	30,581	25.7
1998	14,150	11.9
1999	14,870	12.5
2000	14,902	12.5
2001	15,392	12.9
2002	15,030	12.6
2003	15,093	12.7
2004	14,572	12.3
2005	14,923	12.6
Total	118,932	

Table 5.3 Composition of sample, soldiers vs. civilian employment, health

Demographic characteristic	No. of observations	% sample
Dependent variable		
Soldier	358	1.8
Other	19,887	98.2
Independent variables		
Sex		
Male	10,795	533
Female	9,450	46.7
Germany		
West Germany	14,312	70.7
East Germany	5,933	29.3
Federal state		
Baden-Württemberg	1,336	6.6
Bavaria	2,306	11.4
Berlin	1,003	5.0
Brandenburg	550	2.7
Bremen	600	3.0
Hamburg	1,902	9.4
Hesse	896	4.4
Mecklenburg	1,204	6.0
Lower Saxony	1,194	5.9
North Rhine-Westph.	1,391	6.9
Rhineland-Pfalz	1,271	6.3
Saarland	1,886	9.3
Saxony	1,342	6.6
Saxony-Anhalt	1,065	5.3
Schleswig-Holstein	1,530	7.6
Thuringia	769	3.8

(continued)

Table 5.3 (continued)

Demographic characteristic	No. of observations	% sample
Settlement size		
<20,000 residents	9,123	45.1
>20,000 residents	11,122	54.9
Educational attainment		
Middle level or below	13,665	67.5
Upper level	6,580	32.5
Health		
Normal weight	16,289	80.5
Overweight/Obese	3,956	19.5
Smoking behavior		
Non-smoker	8,794	43.4
Smoker	11,451	56.6
Age group		
19 years	3,983	19.7
20 years	5,031	24.9
21 years	5,554	27.4
22 years	5,677	28.0
Marital status		
Unmarried	19,625	96.9
Married	620	3.1
Per capita household income		
<300 Euros	3,396	16.8
>300 Euros	16,849	83.2
Children in the household		
No child <18 years	15,163	74.9
1+children <18 years	5,082	25.1
1999	5,008	24.7
2003	4,726	23.3
2005	10,511	51.9
Total	20,245	

Table 5.4 Composition of sample, basic conscripts vs. civilian servants

Demographic characteristic	No. of observations	% sample
Dependent variable		
Conscript	2,669	41.7
Other	3,736	58.3
Independent variables		
Germany		
West Germany	4,553	71.1
East Germany	1,852	28.9

(continued)

Table 5.4 (continued)

Demographic characteristic	No. of observations	% sample
Federal state		
Baden-Württemberg	869	13.6
Bavaria	911	14.2
Berlin	250	3.9
Brandenburg	267	4.2
Bremen	53	0.8
Hamburg	155	2.4
Hesse	380	5.9
Mecklenburg	220	3.4
Lower Saxony	524	8.2
North Rhine-Westph.	1,005	15.7
Rhineland-Pfalz	288	4.5
Saarland	126	2.0
Saxony	585	9.1
Saxony-Anhalt	228	3.6
Schleswig-Holstein	242	3.8
Thuringia	302	4.7
Settlement size		
<20,000 residents	3,036	47.4
>20,000 residents	3,369	52.6
Educational attainment		
Middle level or below	3,475	54.3
Upper level	2,930	45.8
Age group		
19 years	1,155	18.0
20 years	2,597	40.6
21 years	1,711	26.7
22 years	942	14.7
Marital status		
Unmarried	6,370	99.5
Married	35	0.6
Per capita household income		
<300 Euros	1,804	28.2
>300 Euros	4,601	71.8
Children in the household		
No child <18 years	4,563	71.2
1 + children <18 years	1,842	28.8
1998	962	15.0
1999	1,045	16.3
2000	931	14.5
2001	947	14.8
2002	855	13.4
2003	641	10.0
2004	567	8.9
2005	457	7.1
Total	6,405	

Table 5.5 Composition of sample, basic conscripts vs. civilian servants, health

Demographic characteristic	No. of observations	% sample
Dependent variable		
Conscript	357	41.3
Other	508	58.7
Independent variables		
Germany		
West Germany	603	69.7
East Germany	262	30.3
Federal state		
Baden-Württemberg	62	7.2
Bavaria	110	12.7
Berlin	39	4.5
Brandenburg	37	4.3
Bremen	28	3.2
Hamburg	55	6.4
Hesse	48	5.6
Mecklenburg	47	5.4
Lower Saxony	55	6.4
North Rhine-Westph.	73	8.4
Rhineland-Pfalz	58	6.7
Saarland	60	6.9
Saxony	61	7.1
Saxony-Anhalt	45	5.2
Schleswig-Holstein	54	6.2
Thuringia	33	3.8
Settlement size		
<20,000 residents	426	49.3
>20,000 residents	439	50.8
Educational attainment		
Middle level or below	470	54.3
Upper level	395	45.7
Health		
Normal weight	712	82.3
Overweight/Obese	153	17.7
Smoking behavior		
Non-smoker	374	43.2
Smoker	491	56.8
Age group		
19 years	173	20.0
20 years	326	37.7
21 years	229	26.5
22 years	137	15.8
Marital status		
Unmarried	861	99.5
Married	4	0.5

(continued)

Table 5.5 (continued)

Demographic characteristic	No. of observations	% sample
Per capita household income		
<300 Euros	222	25.7
>300 Euros	643	74.3
Children in the household		
No child <18 years	626	72.4
1 + children <18 years	239	27.6
1999	344	39.8
2003	171	19.8
2005	350	40.5
Total	865	

employer?". The range of possible answers included: (1) very strongly interested; (2) strongly interested; (3) less interested; (4) not at all interested.

In another question limited to males either not yet mustered or tested fit for military service, respondents were asked: "Do you intend to complete basic military service in the Bundeswehr?". The answer options included: (1) I intend to complete basic military service; (2) I already completed basic military service; (3) I intend to perform alternative civilian community service; (4) I already performed alternative civilian community service; (5) I am undecided.

In an additional sub-question, males that indicated their propensity for basic military service or had already decided for basic military service were asked: "Do you intend to enlist as a regular or professional soldier in the Bundeswehr?". The range of pre-formulated answers included: (1) Yes, I would like to serve as a regular or professional soldier; (2) No, I don't want to serve as a regular or professional soldier; (3) I am undecided; (4) I already serve as a regular or professional soldier. The dependent and independent variables used in the analysis of military propensity are described in more detail in Table 5.6.

5.4.3.2 Military Service

Military service, as the outcome variable of primary interest, is measured by responses to a survey question about current occupational status. In addition to various types of civilian employment, the answer options included "being enlisted as a regular or professional soldier" and "being enlisted as a basic military service conscript". Separate analyses are calculated for the independent variables "regular or professional soldier" and "basic military service conscript". The objective is to not eliminate group-specific differences between professional soldiers vs. others in civilian employment and military service conscripts vs. others performing civilian community service. The dependent and independent variables used in the analysis are described in more detail in Table 5.7.

Table 5.6 Description of the variables, military propensity

Variable	Description	Range
Dependent variables		
Bundeswehr as a potential employer	Identifies individuals that are interested in the Bundeswehr as a potential employer	0 = No interest 1 = Interest
Basic military service	Identifies individuals that intend to enlist as a basic military service conscript in the Bundeswehr	0 = No intention 1 = Intention
Military service as a regular or professional soldier	Identifies individuals that intend to serve as a basic military service conscript (Filter, see above) and also intend to enlist as a regular or professional soldier in the Bundeswehr at a later date	0 = No intention 1 = Intention
Independent variables		
Sex	Males, females	0 = male (reference) 1 = female
West Germany/East Germany	Identifies individuals living in the former territory of the Federal Republic (West) or territory of the former GDR (East).	0 = West (reference) 1 = East
Settlement size	Classifies the size of individuals' places of residence into communities with fewer than 20,000 inhabitants and with more than 20,000 residents.	0 = <20,000 inhabitants (reference) 1 = >20,000 residents
Educational attainment	Identifies the current educational status of individuals as middle level education or below (including lower secondary education and apprenticeships) or upper level education (including upper secondary, post secondary, and tertiary education).	0 = middle level education or below (reference) 1 = upper level education
Age	Grouped	16–18 years (reference) 19–20 years

Source: Youth opinion poll on the security political situation in Germany (2006)

5.4.4 Modeling Strategy

Model specification and the order, in which independent variables were sequentially entered into the model, were based on an initial test of association between each of the independent variables and the respective outcome variable of military interest or military service, as suggested by Hosmer and Lemeshow (2000: 106), and theoretical considerations. Hence, in a first step, the effect of sex as the main distinctive feature of military propensity and military service was estimated. Then the regional origin was entered (i.e. East vs. West Germany or federal state) and an indicator of urbanity (i.e. settlement size) in order to control for contextual differences in general socio-economic conditions and the availability of employment

Table 5.7 Description of the variables, military service

Variable	Description	Range
Dependent variables		
Basic military service conscript	Identifies individuals as undergoing basic military training vs. performing alternative civilian community service.	0 = civilian community service 1 = basic military service in the Bundeswehr
Regular or professional soldier	Identifies individuals as being enlisted in the military as a regular or professional soldier vs. pursuing other civilian employment.	0 = other (self-employed, family worker, civil servant, judge, employee, worker, home worker, trades/technical trainee, commercial trainee) 1 = regular or professional soldier in the Bundeswehr (also including Federal Police and riot police)
Independent variables		
Sex	Males, females	0 = male (reference) 1 = female
Regional origin		
West Germany/ East Germany	Identifies individuals living in the former territory of the Federal Republic (West) or territory of the former GDR (East).	0 = West (reference) 1 = East
Federal state	Identifies individuals living in Bavaria, Baden-Württemberg, Berlin, Brandenburg, Bremen, Hamburg, Hesse, Lower Saxony, Mecklenburg- Western Pomerania, North Rhine-Westphalia, Rhineland-Pfalz, Saarland, Saxony, Saxony-Anhalt, Schleswig-Holstein, Thuringia.	1–16 (Baden-Wuerttemberg (8) = reference for West) (Saxony (14) = reference for East)
Settlement size	Classifies the size of individuals' places of residence into communities with fewer than 20,000 inhabitants and with more than 20,000 residents.	0 = <20,000 inhabitants (reference) 1 = >20,000 residents
Socio-demographic traits		
Educational attainment	Identifies the highest educational attainment of individuals as middle level education or below (i.e. "Realschulabschluss" or "Hauptschulabschluss") or upper level education (i.e. "Abitur").	0 = middle level education or below (reference) 1 = upper level education
Age	Raw value	19–22 (youngest age = reference)
Marital status	Unmarried, married	0 = unmarried (reference) 1 = married

(continued)

Table 5.7 (continued)

Variable	Description	Range
Health status		
Body mass index (BMI)	Assesses individuals as being normal weight or overweight/obese. Calculated as weight in kg, divided by height in meters squared.	0 = normal weight (BMI >= 18.5 and < 25) (reference) 1 = overweight/obese (BMI >= 25)
Smoking behavior	Identifies individuals as non-smoker (including ex-smokers) and smoker.	0 = non-smoker (reference) 1 = smoker
Family background		
Per capital household income	Monthly household per capita income is defined as the total monthly income of household members over the total number of household members.	0 = <300 Euro 1 = >300 Euro (reference)
Siblings <18 years	Informs about the number of children below 18 years in the household, i.e. the number of younger siblings of the 19–22 year-old population at risk.	0 = no siblings (reference) 1 = one or more siblings
Time	Year	1998–2005 (first year = reference)

Source: Scientific Use Files of the German Microcensus 1998–2005

opportunities. The reference categories of the Western and Eastern group of federal states (Baden-Wuerttemberg and Saxony) both feature the greatest number of observations and a comparatively favorable economic outlook.

Human capital is measured by two aspects: Firstly, respondents' highest level of educational attainment serves as a proxy for learning proficiency. Secondly, the analysis of the likelihood of military service takes advantage of the irregularly acquired health information regarding respondents' body composition and smoking behavior. Body composition, as specified by the body mass index, and smoking behavior have been found to be strong predictors for overall health status. School dropouts are not considered in the analysis. Without the successful completion of a lower level course of education, they do not fulfill the minimal requirement for military entry and therefore do not belong to the population at risk. For the same reason, all individuals with non-German citizenship are excluded.

An additional age variable is included to demonstrate the effect of age-specific graduation rates. In the analysis of actual military service, there is also a control for family characteristics that may interfere with military service, such as, the respondent's marital status, and in terms of the family of origin, the level of per capita household income, and the presence of younger children (i.e. siblings) in the household. Lastly, with respect to the analysis of actual military service, time in calendar years is included as a proxy for the military's institutional change and associated force reductions, in particular among basic military conscripts.

Fig. 5.2 Overview of regression models, military vs. civilian employment

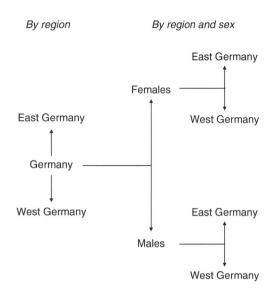

In consideration of the "principle of parsimony in modeling" (Hosmer and Lemeshow 2000: 191), as few parameters as possible are used and dichotomized. Thus, the complexity of the previously multi-category covariates settlement size and educational attainment, body composition, smoking behavior, marital status, per capita household income, and children in the household is reduced in order to secure a relevant effect size and improve the goodness of the model fit. While these statistical considerations may also motivate to dichotomize the age variable in the analysis of enlistment propensity, the multi-level coding for age and year is specifically kept to demonstrate potential differences over age and time.

In view of the well-documented socio-demographic differences in military propensity and military service between youth in East and West Germany, separate regression analyses, always data permitting, are run for the German youth population at risk, then for youth from East or West Germany, and for males and females. One series of models is estimated with the pooled Microcensus data from 1998 to 2005 that disregards health status. Another series of models is calculated on the basis of the pooled Microcensus data from 1999, 2003 and 2005 taking advantage of the irregular additional survey program on respondents' health status. Due to the small sample size, no logistic regression models with independent health variables can be estimated for female soldiers from East or West Germany.

Figure 5.2 provides an overview of the estimated logistic regression models predicting military service as a regular or professional soldier relative to pursuing other civilian employment. An additional set of logistic regression models predicting military service as a basic military conscript relative to performing alternative civilian community service is estimated for the German youth population of males, and separately, for males originating from East or West Germany.

While regression models slightly vary according to the respective subpopulation studied, the general regression equation to calculate the predicted outcome value of military service for an individual with given characteristics on the independent variables, can be expressed as follows:

$$odds = \frac{1}{1-p} = e^{b_0} \times e^{b_1 X_1} \times e^{b_2 X_2} \times e^{b_3 X_3} \times e^{b_4 X_4} \times e^{b_5 X_5} \times e^{b_6 X_6} \times$$
$$e^{b_7 X_7} \times e^{b_8 X_8} \times e^{b_9 X_9} \times e^{b_{10} X_{10}} \times e^{b_{11} X_{11}}$$

where

$\dfrac{1}{1-p}$ = predicted odds ratio on the outcome measure

e^{b_0} = intercept

$e^{b_1 X_1}$ = sex

$e^{b_2 X_2}$ = regional origin

$e^{b_3 X_3}$ = settlement size

$e^{b_4 X_4}$ = educational attainment

$e^{b_5 X_5}$ = body composition

$e^{b_6 X_6}$ = smoking behavior

$e^{b_7 X_7}$ = age

$e^{b_8 X_8}$ = marital status

$e^{b_9 X_9}$ = per capita household income

$e^{b_{10} X_{10}}$ = presence of younger siblings in the household

$e^{b_{11} X_{11}}$ = year

In order to preclude bias from multicollinearity, i.e. the condition of highly correlated independent variables in the model, uncentered variance inflation factors (VIFs) were computed and reviewed for all independent variables. However, the estimated correlation coefficients did not contain a reference to collinearities among the independent variables and therefore did not influence model specification.

The nature and strength of the association between the independent variables and the outcome military service are described by means of odds ratios (OR). They represent an alternative expression of the logistic regression coefficient β and result from the exponentiation of the logit coefficient, calculating $odds(Y=1)=e^{\log it(Y)}$, while still expressing exactly the same relationship (Menard 1995: 12). Odds ratios are a form of relative difference or ratio and have found wide application in the social sciences. They are usually the parameter of interest in a logistic regression, since they provide an easily interpreted measure of how much more likely (or unlikely) it is for an outcome to be present among respondents that display a certain attribute, like being married, than among those without it, thus being unmarried (Hosmer and Lemeshow 2000: 50).

The interpretation of odds ratios can be illustrated as follows: in one regression model of the chances of German youth to serve in the military, the antilog of one logistic regression coefficient β is $e^{\beta}=1,249$. That means the odds of military service for a subject with younger siblings are estimated to be 1.249 times larger than the odds for a subject who lives without younger children in the household. Hence, the odds increase by about 25 %. In the same model, the antilog of logistic regression coefficient β for educational attainment is 0.709, indicating that a subject with an upper level education is approximately 0.709 times as likely to serve in the military as a similar subject of a middle level educational background or below. Thus, if an odds ratio for an individual of certain socio-demographic attributes is less than one, then the outcome is less likely to occur for this individual in comparison with individuals that fall into the defined reference category.

For each independent variable in the regression models, p-values were generated to provide information about whether a variable is statistically significantly associated with the outcome. Three levels of significance, i.e. probability of error, were differentiated and denoted $p < 0.01$, $p < 0.05$, and $p < 0.1$.[5] Effects without such symbols are not statistically significant. In addition, logistic regression models were compared on the basis of the likelihood ratio test and the difference in the values of $(-2 \log L)$ for two consecutive models with varying sets of predictor variables. This measures provides an approximate chi-squared statistic of independence with the degrees of freedom (df) indicating the number of extra parameters in the model (Agresti and Finlay 1999: 584). To illustrate, one might return to the regression model of German youth: the initial model containing only an intercept has $(-2 \log L) = -10,304.085$. In the full model, after adding all available socio-demographic background factors, the $(-2 \log L)$ increased to $-9,088.143$. This difference is reflected in the likelihood ratio (LR) chi-square test statistic, which represents minus two times the difference of intercept model and the full model, that means $-2 \times (10,304.085 - 9,088.143) = 1,963$ with degrees of freedom $= 17$.

McFadden's pseudo-R2 value was used as a supplemental measure of explanatory power of the statistical models (Backhaus et al. 2000: 116).[6] It is a popular measure of fit in statistical modeling and based on a comparison of the log likelihood of the full and intercept model:

$$\text{McFadden's R-squared} = 1 - \frac{LL_{Full}}{LL_0}$$

If the difference between the full and intercept model is small, the quotient is close to unity and, as a result, the McFadden's R-squared close to zero. As more predictors are added, McFadden's R-squared may move towards unity, thereby indicating a good model fit. Yet Backhaus et al. (2000: 116) suggest that parameter values between 0.2 and 0.4 already contain a reference to an appropriate model fit. All calculations were carried out with the statistical software package STATA version 10.

---

[5] The levels of significance are denoted as follows: *** $p < 0.01$; ** $p < 0.05$; * $p < 0.1$.

[6] The McFadden's pseudo R-squared is an attempt to compensate for the lack of an unambiguous goodness-of-fit measure like the R-squared in linear regression that informs about the fraction of the total variance of the outcome variable that is "explained" by variation in the independent variables. However, the specification of the McFadden's pseudo R-squared is not without controversy.

5.5 Results

The present section reports the results of the multivariate analyses. Models of military propensity and actual military service were examined using logistic regression, taking into consideration key socio-demographic characteristics. MLE regression was applied to predict the probability of military propensity and military service among German youth, and where applicable, a set of defined subpopulations. Only estimation results of each full model are represented and, for reasons of clarity, results of preceding preliminary main effects models are discounted, in which the net effect of each explanatory variable was examined one by one (Hosmer and Lemeshow 2000: 97).

5.5.1 Military Propensity

The results of the multivariate logistic regression models predicting interest in military service are presented in Table 5.8. Separate analyses were conducted regarding youth's interest in the Bundeswehr as a potential employer (Model I), as well as males' intentions regarding basic military service (Model II) and a military career as a regular or professional soldier (Model III).

Turning to Model I, which examines the socio-demographic determinants of military propensity for males and females alike, females are, as expected, less likely to be interested in the Bundeswehr as a potential employer. However, the relatively narrow gender difference in the prediction of propensity is somewhat surprising: females are only about 20 % less likely than males to consider the Bundeswehr as a potential place of employment (OR = 0.773). This circumstance, however, may stem from the fact that the respective survey question did not differentiate between non-combat and combat-related functions.

When considering all three regression models at large, the pattern of estimation results is overwhelmingly consistent. Residence in East Germany turns out to be a strong and positive predictor of military propensity (e.g. OR = 1.636 in Model I), while a higher level of urbanity is found to reduce the likelihood of being interested in any kind of military career (e.g. OR = 0.825 in Model I).

In equal measure, there was a consistently negative association between respondents' level of educational attainment and their interest in the Bundeswehr as a potential employer (OR = 0.614 in Model I) or any type of military employment. This negative association is particularly pronounced when males are asked about their interest in basic military service relative to civilian community service (OR = 0.462 in Model II).

With regard to the influence of age on military propensity, there is a consistent, yet statistically insignificant, association. The size of the negative effect is stronger for males' interest in basic military service (OR = 0.686 in Model II) and serving as professional or regular soldier (OR = 0.686 in Model III) than for youth's interest in the Bundeswehr as a potential employer (OR = 0.947 in Model I).

Table 5.8 Results, interest in military service (OR), Germany, males/females

	Model I		Model II		Model III	
	Interest in the Bundeswehr as a potential employer		Interest in basic military service		Interest in serving as a regular or professional soldier	
	OR	p	OR	p	OR	p
Male	1					
Female	0.773	***				
Germany			Germany		Germany	
West	1		1		1	
East	1.636	***	1.624	***	1.383	*
Settlement size			Settlement size		Settlement size	
<20,000 residents	1		1		1	
>20,000 residents	0.825	*	0.901		0.802	
Educational attainment			Educational attainment		Educational attainment	
Middle level or below	1		1		1	
Upper level	0.614	***	0.462	***	0.636	
Age group			Age group		Age group	
16–18 years	1		1		1	
19–20 years	0.947		0.686	*	0.686	
Measures of model fit			*Measures of model fit*		*Measures of model fit*	
No. of observations	1,745		635		285	
Log-likelihood (0)	−1,153		−440		−191	
Log-likelihood	−1,127		−423		−188	
LR [=LR chi2]	53		33		6	
McFadden's R2	0.02		0.04		0.02	
df	8		11		6	

5.5.2 Military Service

5.5.2.1 Service as a Regular or Professional Soldier

The odds ratios and significance levels obtained from the first series of regression models are presented in Table 5.9. They provide information about the socio-demographic determinants of military service among youth in Germany. The results show that sex acts as the main determinant of military service. Regardless of whether the aggregate youth population in Germany (Model 1) is concerned, or whether East and West Germany (Models 2 and 3) are considered individually, the chances of a female subject to serve as a regular or professional soldier are negligible and universally range below 6 % of those for a male subject with similar socio-demographic traits.

In a closer analysis of local characteristics, it appears that residence in East Germany has a strong positive impact on the likelihood of an individual to choose the military as a profession. When the aggregate youth population of Germany is concerned (Model 1), the odds of an East German subject to serve in the military were nearly 80 % higher than those of a similar subject from West Germany (OR = 1.756). On the federal state level and with respect to East Germany (Model 2), young males and females living in Mecklenburg (OR = 1.575) and Saxony-Anhalt (OR = 1.411) are more likely to serve in the military relative to the reference state Saxony. In West Germany (Model 3), the youth population of Bremen (OR = 2.463), Schleswig-Holstein (OR = 2.234), and Lower Saxony (OR = 1.856) is about twice as likely to serve in the military as their peers in the reference state Baden-Württemberg.

Evidence of the effect of a respondent's residence in a city state like Berlin, Bremen or Hamburg supports the hypothesis that urban living reduces the likelihood of military service. A subject that resides in Berlin is significantly less likely to serve in the military (OR = 0.698) than a similar subject from the reference state Saxony (Model 2). Yet, subjects from Hamburg and especially Bremen show a high likelihood to serve in the military and thereby contradict the prediction about the negative effect of urbanity (Model 3).

Turning to the effect of settlement size as a proxy for the level of urbanity, results for the youth population in Germany show that residence in a settlement of more than 20,000 residents is associated with a 30 % lower likelihood to serve in the military relative to a subject that lives in a smaller town (OR = 0.709 in Model 1). This inverse relationship between urbanity and military service equally shows in the two less aggregated models of East Germany (Model 2) and West Germany (Model 3), while in the former, the effect is less pronounced (OR = 0.826).

A differentiation of youth by major region reveals that educational attainment acts differently on the likelihood of an individual from East Germany and West Germany. In East Germany (Model 2), a subject with a high educational attainment is about 30 % *more* likely to serve in the military than a similar subject with a low or middle level education (OR = 1.304). In West Germany (Model 3), educational

Table 5.9 Results, regular soldier (OR), Germany, males/females

Model 1

Germany	OR	p
Male	1	
Female	0.056	***
Germany		
West	1	
East	1.756	***
Settlement size		
<20,000 residents	1	
>20,000 residents	0.709	***
Educational attainment		
Middle level or below	1	
Upper level	0.941	***
Age group		
19 years	1	
20 years	1.719	***
21 years	2.250	***
22 years	2.321	***

Model 2

East Germany	OR	p
Male	1	
Female	0.059	***
Federal state		
Berlin	0.698	***
Brandenburg	1.119	
Mecklenburg	1.575	***
Saxony	1	
Saxony-Anhalt	1.411	***
Thuringia	1.206	
Settlement size		
<20,000 residents	1	
>20,000 residents	0.826	***
Educational attainment		
Middle level or below	1	
Upper level	1.304	***
Age group		
19 years	1	
20 years	1.960	***
21 years	2.386	***
22 years	2.364	***

Model 3

West Germany	OR	p
Male	1	
Female	0.054	***
Federal state		
Baden-Württemberg	1	
Bavaria	1.292	***
Bremen	2.463	***
Hamburg	1.440	**
Hesse	1.314	**
Lower Saxony	1.856	***
North Rhine-Westph.	0.973	
Rhineland-Pfalz	1.368	**
Saarland	1.559	**
Schleswig-Holstein	2.234	***
Settlement size		
<20,000 residents	1	
>20,000 residents	0.728	***
Educational attainment		
Middle level or below	1	
Upper level	0.772	***
Age group		
19 years	1	
20 years	1.556	***
21 years	2.127	***
22 years	2.273	***

	Model 1		Model 2		Model 3	
Marital status						
Unmarried	1		1		1	
Married	0.693	**	0.511		0.756	
Per capita household income						
<300 Euros	0.520	***	0.292	***	0.793	***
>300 Euros	1		1		1	
Children in the household						
No child <18 years	1		1		1	
1+ children <18 years	1.249	***	1.512	***	1.116	*
1998	1		1		1	
1999	1.045		1.217		0.945	
2000	0.852	*	0.915		0.821	*
2001	0.921		1.068		0.844	*
2002	0.876		1.015		0.805	*
2003	0.926		0.990		0.901	
2004	0.851	*	0.938		0.814	*
2005	0.714	***	0.814		0.502	***
Measures of model fit						
No. of observations	118,932		33,840		85,092	
Log-likelihood (0)	−10,304		−3,849		−6,390	
Log-likelihood	−9,088		−3,328		−5,661	
LR [=LR chi2]	2,432		1,042		1,458	
McFadden's R2	0.12		0.14		0.11	
df	17		21		25	

attainment has a detrimental effect on the likelihood of military service. More educated subjects are about 20 % *less* likely to serve in the military than subjects with a lower educational attainment (OR = 0.772).

The effect of age on the likelihood of military service is universally positive, at least when the age range of 19–22 years is under observation. Accordingly, the odds of military service are more than two times higher for a 22-year-old relative to a 19-year-old with similar socio-demographic characteristics (OR = 2.321 in Model 1). As for marital status, the results seem to confirm the hypothesis that marriage reduces the appeal of military service. When the aggregate youth population is concerned (Model 1), the likelihood of military service among subjects that are already married at 19–22 years is only about 70 % that of their unmarried peers (OR = 0.693). In East Germany (Model 2), the negative effect of being married on military service tends to be even larger (OR = 0.511).

Conversely, the initial prediction about the influence of the family background on the likelihood of military service only finds limited affirmation in the data. For example, contrary to the formulated expectation, a subject from a family with less than 300 Euros of per capita household income is *less* likely to serve in the military than a subject from a more affluent family background. In Germany (Model 1), being from a family with a low level of disposable income for each individual cuts the chances of military service by nearly half (OR = 0.520). This effect is even more pronounced in East Germany (Model 2), where a low per capita household income results in a likelihood to serve in the military that is only 30 % that of a subject from a more affluent family (OR = 0.292). On the other hand, the presence of children below 18 years in the respondent's household, i.e. younger siblings, increases the likelihood of military service in the aggregate youth population by about 25 % (Model 1), and even more so in East Germany, where the odds of military service for a subject with siblings are about 50 % higher than those of a lone child (Model 2).

As predicted, the likelihood of military service varies inversely with time. Hence, the probability of military service was consistently lower in the years subsequent to the reference year 1998. Especially in West Germany, the likelihood of military service declined over time. For example, a subject in 2005 was only half as likely to serve in the military as a similar subject in 1998 (Model 3).

Under the assumption that there are sex-specific differences in the effect of socio-demographic attributes on the likelihood of military service as a regular or professional soldier, two separate sets of regression models are estimated for the male and female youth population in Germany, as well as in East and West Germany. The results for the male population, which are reported in Table 5.10, confirm the size and the direction of effects already observed in the aggregate model for both males and females. Thus, for males in Germany (Model 4), important predictors of military service include the residence in East Germany (OR = 1.737), membership in an advanced age group (OR = 2.480 for a 22-year-old), and the affiliation with a household with children below 18 years (OR = 1.283). Conversely, the results confirm a range of factors with a detrimental effect on the likelihood of military service among males, such as living in a settlement of more than 20,000 people (OR = 0.717), having completed an upper level school education (OR = 0.906), being married

Table 5.10 Results, regular soldier (OR), Germany, males

Model 4			Model 5			Model 6		
Germany, males	OR	p	East Germany, males	OR	p	West Germany, males	OR	p
Germany			Federal state			Federal state		
West	1		Berlin	0.705	**	Baden-Württemberg	1	
East	1.737	***	Brandenburg	1.103		Bavaria	1.251	**
			Mecklenburg	1.542	***	Bremen	2.344	***
			Saxony	1		Hamburg	1.341	
			Saxony-Anhalt	1.370	***	Hesse	1.226	***
			Thuringia	1.144		Lower Saxony	1.807	***
						North Rhine-Westph.	0.964	
						Rhineland-Pfalz	1.311	**
						Saarland	1.573	**
						Schleswig-Holstein	2.234	***
Settlement size			Settlement size			Settlement size		
<20,000 residents	1		<20,000 residents	1		<20,000 residents	1	
>20,000 residents	0.717	***	>20,000 residents	0.835	**	>20,000 residents	0.732	***
Educational attainment			Educational attainment			Educational attainment		
Middle level or below	1		Middle level or below	1		Middle level or below	1	
Upper level	0.906	*	Upper level	1.265	***	Upper level	0.740	***
Age group			Age group			Age group		
19 years	1		19 years	1		19 years	1	
20 years	1.786	***	20 years	2.093	***	20 years	1.590	***
21 years	2.379	***	21 years	2.626	***	21 years	2.189	***
22 years	2.480	***	22 years	2.620	***	22 years	2.372	***
Marital status			Marital status			Marital status		
Unmarried	1		Unmarried	1		Unmarried	1	
Married	0.600	**	Married	0.608		Married	0.615	**

(continued)

Table 5.10 (continued)

	Model 4 Germany, males		Model 5 East Germany, males		Model 6 West Germany, males	
	OR	p	OR	p	OR	p
Per capita household income						
<300 Euros	0.546	***	0.311	***	0.823	**
>300 Euros	1		1		1	
Children in the household						
No child <18 years	1		1		1	
1+ children <18 years	1.283	***	1.587	***	1.132	*
Time						
1998	1		1		1	
1999	1.050		1.267		0.926	
2000	0.874		0.932		0.847	
2001	0.921		1.107		0.824	*
2002	0.866		1.010		0.793	**
2003	0.912		0.961		0.896	
2004	0.837	*	0.933		0.796	**
2005	0.698	***	0.796		0.492	***
Measures of model fit						
No. of observations	60,464		17,645		42,819	
Log-likelihood (0)	−8.554		−3.174		−5.326	
Log-likelihood	−8.326		−3.031		−5.203	
LR [=LR chi2]	455		286		247	
McFadden's R2	0.03		0.05		0.02	
df	16		20		24	

(OR = 0.600), having a low per capita household income (OR = 0.546), and time (OR = 0.698 for 2005).

By and large, these determining factors also become evident in separate regression models for the male population in East and West Germany (Models 5 and 6). However, specific differences in the range and, when education is concerned, in the direction of effects confirm the relative East–west divide in the socio-demographic determinants of military service already observed for the aggregate youth population of males and females in Models 2 and 3. Thus, at the level of federal states, it is again residence in economically weaker regions that is associated with a higher likelihood of military service. Males from Mecklenburg (OR = 1.542) and Saxony-Anhalt (OR = 1.370) have the highest odds of military service in East Germany (Model 5), while in West Germany, males from Bremen (OR = 2.344), Schleswig-Holstein (OR = 2.234) and Lower Saxony (OR = 1.807) are most likely to serve in the military (Model 6).

With respect to the influence of human capital on the likelihood of military service, the previous finding about the opposite effect of educational attainment in East and West Germany is substantiated (cf. Models 2 and 3). Accordingly, better educated males in East Germany are about 26 % *more* likely to serve in the military than males with a middle level education or less (OR = 1.265 in Model 5). Estimation results for males in West Germany reaffirm the predicted negative association between an individual's educational attainment and likelihood to serve. There, a male subject with an upper level education is estimated to be about 25 % *less* likely to serve as regular or professional soldier (OR = 0.740 in Model 6).

Turning to the estimation results for the female population as shown in Table 5.11, the marked influence of local characteristics on the likelihood of military service reappears. A female from East Germany is about twice as likely to serve in the military as a similar subject from West Germany (OR = 2.101 in Model 7). Residence in less favorable regions, in view of the local labor market situation, is associated with a greater likelihood of females to join the military. In addition to the established relationship between military service and residence in the East German states of Mecklenburg or Saxony-Anhalt, as well as the West German states of Schleswig Holstein, Bremen and Lower Saxony (cf. Models 2 and 3), females residing in Thuringia (OR = 2.871), Hamburg (OR = 4.471) and Hesse (OR = 4.471) are much more likely to serve than their peers in the respective reference states.

In contrast to the initial prediction and the results estimated for the male population, there is a consistently positive association between females' educational attainment and military service. Regardless of whether the female population in Germany or only in East or West Germany is concerned, more highly educated females generally have a higher likelihood to serve in the military than their less educated peers. This positive association is particularly pronounced among females in East Germany (OR = 1.940 in Model 8).

Similarly, the influence of family background variables on females' likelihood to serve in the military evolves in a fairly unexpected way and does not reflect the predictive association established for the male population. Noteworthy is the strong negative impact of low per capita household income on females' likelihood to serve

Table 5.11 Results, regular soldier (OR), Germany, females

Model 7			Model 8			Model 9		
Germany, females	OR	p	East Germany, females	OR	p	West Germany, females	OR	p
Germany			Federal state			Federal state		
West	1		Berlin	0.623		Baden-Württemberg	1	
East	2.101	***	Brandenburg	1.487		Bavaria	2.693	*
			Mecklenburg	2.390	*	Bremen	6.150	**
			Saxony	1		Hamburg	4.471	**
			Saxony-Anhalt	2.430	*	Hesse	4.386	***
			Thuringia	2.871	**	Lower Saxony	3.541	**
						North Rhine-Westph.	1.263	
						Rhineland-Pfalz	3.273	*
						Saarland	1.232	
						Schleswig-Holstein	2.262	
Settlement size			Settlement size			Settlement size		
<20,000 residents	1		<20,000 residents	1		<20,000 residents	1	
>20,000 residents	0.572	***	>20,000 residents	0.676		>20,000 residents	0.679	
Educational attainment			Educational attainment			Educational attainment		
Middle level or below	1		Middle level or below	1		Middle level or below	1	
Upper level	1.644	**	Upper level	1.940	**	Upper level	1.369	
Age group			Age group			Age group		
19 years	1		19 years	1		19 years	1	
20 years	0.949		20 years	0.897		20 years	1.079	
21 years	0.936		21 years	0.667		21 years	1.362	
22 years	0.803		22 years	0.594		22 years	1.181	

	Model 1	Model 2	Model 3
Marital status			
Unmarried	1		
Married	1.950 *		
Per capita household income			
<300 Euros	0.146 ***	0.052 ***	0.293 **
>300 Euros	1	1	1
Children in the household			
No child <18 years	1	1	1
1+ children <18 years	0.760 *	0.616	0.964 *
1998	1	1	1
1999	0.975	0.486	1.364
2000	0.442	0.641	0.268
2001	0.921	0.465	1.304
2002	1.064	1.052	1.057
2003	1.180	1.384	1.007
2004	1.137	1.025	1.188
2005	1.040	1.029	0.727
Measures of model fit			
No. of observations	58,468	16,207	42,347
Log-likelihood (0)	–769	–304	–460
Log-likelihood	–742	–281	–443
LR [=LR chi2]	53	46	33
McFadden's R2	0.03	0.08	0.04
df	16	19	23

in Germany (OR = 0.146 in Model 7) and East Germany (OR = 0.052 in Model 8). Equally unanticipated is the consistent negative effect of younger dependents on females' likelihood to serve in the military (OR = 0.760 in Model 7). Then again, the opening of all military career tracks for women in 2001 is reflected in their somewhat higher likelihood to serve towards the end of the observation period, especially in East Germany (cf. Models 7 and 8).

Turning to the association between health status and military service displayed in Table 5.12, the estimation results for youth in Germany prove that overweight or obese individuals are consistently less likely to serve in the military than individuals of normal weight (e.g. OR = 0.862 in Model 10).[7] Also in line with the initial expectation, there is no predictive association between smoking behavior and military service among the general youth population in Germany (e.g. OR = 1.079 in Model 10). If males in East and West Germany are concerned (see Table 5.13), the direction and the size of the health effect on military service are almost identical with the estimation results for all Germany. That means, overweight or obese males have lower chances to serve in the military than their normal weight counterparts (e.g. OR = 0.878 in Model 13), while the predictive relationship between smoking behavior and military serve is rather weak (e.g. OR = 1.068 in Model 13).

However, when the influence of health factors on military service is reviewed separately for males and females in order to account for sex-specific differences in body composition and health-related behaviors (see Table 5.14), it turns out that only the estimation results for males are in line with the initially formulated prediction (Model 15). In contrast, overweight or obese females are almost twice as likely to serve in the military as their normal weight counterparts (OR = 1.860 in Model 16), while female smokers are somewhat less likely to serve than female non-smokers (OR = 0.903 in Model 16).

5.5.2.2 Service as a Basic Military Conscript

The second series of multivariate analyses examines socio-demographic predictors of undergoing basic military training relative to performing alternative civilian community service (see Table 5.15). It is striking that only the size and the direction of the effect of local characteristics on military service resemble the predictive associations established for regular or professional soldiers in the models described above.[8] Accordingly, males from East Germany are almost twice as likely to undergo basic military training as their peers from West Germany (OR = 1.934 in Model 17). Looking at East and West Germany individually, it appears that the likelihood of basic military training among males from Mecklenburg is more than 150 % higher

---

[7] Given that all health-related models are based on a reduced sample size, the reliability of our findings is generally limited due to statistical insignificance.

[8] The terms "basic military training" or "basic military service" are used interchangeably to denote the German term "Wehrdienst". The equally common translation "compulsory military service" is not considered appropriate given the great number of exemptions from military service.

Table 5.12 Results, regular soldier (OR), Germany, males/females, health

Model 10

Germany	OR	p
Male	1	
Female	0.064	***
Germany		
West	1	
East	2.041	***
Settlement size		
<20,000 residents	1	
>20,000 residents	0.716	***
Educational attainment		
Middle level or below	1	
Upper level	1.129	
Health		
Normal weight	1	
Overweight/Obese	0.862	
Smoking behavior		
Non-smoker	1	
Smoker	1.079	

Model 11

East Germany	OR	p
Male	1	
Female	0.066	***
Federal state		
Berlin	0.537	**
Brandenburg	0.872	
Mecklenburg	0.756	
Saxony	1	
Saxony-Anhalt	0.877	
Thuringia	0.935	
Settlement size		
<20,000 residents	1	
>20,000 residents	0.743	*
Educational attainment		
Middle level or below	1	
Upper level	1.436	*
Health		
Normal weight	1	
Overweight/Obese	0.945	
Smoking behavior		
Non-smoker	1	
Smoker	1.032	

Model 12

West Germany	OR	p
Male	1	
Female	0.060	***
Federal state		
Baden-Württemberg	1	
Bavaria	0.514	**
Bremen	1.829	*
Hamburg	0.772	
Hesse	1.101	
Lower Saxony	1.325	
North Rhine-Westph.	0.495	**
Rhineland-Pfalz	0.992	
Saarland	0.588	
Schleswig-Holstein	0.594	
Settlement size		
<20,000 residents	1	
>20,000 residents	0.748	*
Educational attainment		
Middle level or below	1	
Upper level	0.896	
Health		
Normal weight	1	
Overweight/Obese	0.786	
Smoking behavior		
Non-smoker	1	
Smoker	1.109	

(continued)

Table 5.12 (continued)

	Model 10 Germany			Model 11 East Germany			Model 12 West Germany		
	OR	p		OR	p		OR	p
Age group									
19 years	1			1			1	*
20 years	0.758			0.802			0.706	
21 years	1.204			1.080			1.288	
22 years	1.481	***		1.172			1.776	***
Marital status									
Unmarried	1						1	
Married	0.574	***					0.890	
Per capita household income									
<300 Euros	0.406	***		0.196	***		0.687	
>300 Euros	1			1			1	
Children in the household									
No child <18 years	1			1			1	
1+children <18 years	1.323	**		1.101			1.546	***
1999	1			1			1	
2003	0.968			0.758			1.239	
2005	0.720	***		0.711	*		0.815	
Measures of model fit									
No. of observations	20,245			5,933			14,312	
Log-likelihood (0)	−1.799			−747			−1.031	
Log-likelihood	−1.589			−653			−910	
LR [=LR chi2]	421			188			242	
McFadden's R2	0.12			0.13			0.12	
df	14			17			22	

Table 5.13 Results, regular soldier (OR), Germany, males, health

Model 13			Model 14		
East Germany, males	OR	p	West Germany, males	OR	p
Federal state			Federal state		
Berlin	0.564	**	Baden-Württemberg	1	
Brandenburg	0.770		Bavaria	0.563	*
Mecklenburg	0.739		Bremen	1.743	
Saxony	1		Hamburg	0.805	
Saxony-Anhalt	0.805		Hesse	1.065	
Thuringia	0.931		Lower Saxony	1.278	
			North Rhine-Westph.	0.545	*
			Rhineland-Pfalz	1.079	
			Saarland	0.646	
			Schleswig-Holstein	0.645	
Settlement size			Settlement size		
<20,000 residents	1		<20,000 residents	1	
>20,000 residents	0.771		>20,000 residents	0.730	*
Educational attainment			Educational attainment		
Middle level or below	1		Middle level or below	1	
Upper level	1.368		Upper level	0.934	
Health			Health		
Normal weight	1		Normal weight	1	
Overweight/Obese	0.878		Overweight/Obese	0.774	
Smoking behavior			Smoking behavior		
Non-smoker			Non-smoker	1	
Smoker	1.068		Smoker	1.102	
Age group			Age group		
19 years	1		19 years	1	
20 years	0.762		20 years	0.702	
21 years	1.176		21 years	1.331	
22 years	1.243		22 years	1.810	***
			Marital status		
			Unmarried	1	
			Married	0.778	
Per capita household income			Per capita household income		
<300 Euros	0.210	***	<300 Euros	0.685	
>300 Euros	1		>300 Euros	1	
Children in the household			Children in the household		
No child <18 years	1		No child <18 years	1	
1+ children <18 years	1.174		1+ children <18 years	1.595	***
1999	1		1999	1	
2003	0.695		2003	1.305	
2005	0.631		2005	0.834	
Measures of model fit			*Measures of model fit*		
No. of observations	3,231		No. of observations	7,564	
Log-likelihood (0)	−619		Log-likelihood (0)	−869	
Log-likelihood	−593		Log-likelihood	−839	
LR [=LR chi2]	52		LR [=LR chi2]	61	
McFadden's R2	0.04		McFadden's R2	0.04	
df	16		df	21	

Table 5.14 Results, regular soldier (OR), Germany, males/females, health

Model 15			Model 16		
Germany, males	OR	p	Germany, females	OR	p
Germany			Germany		
West	1		West	1	
East	2.021	***	East	2.432	*
Settlement size			Settlement size		
<20,000 residents	1		<20,000 residents	1	
>20,000 residents	0.722	***	>20,000 residents	0.603	
Educational attainment			Educational attainment		
Middle level or below	1		Middle level or below	1	
Upper level	1.126		Upper level	1.180	
Health			Health		
Normal weight	1		Normal weight	1	
Overweight/Obese	0.823		Overweight/Obese	1.860	
Smoking behavior			Smoking behavior		
Non-smoker	1		Non-smoker	1	
Smoker	1.089		Smoker	0.903	
Age group			Age group		
19 years	1	*	19 years	1	
20 years	0.736		20 years	1.000	
21 years	1.272	***	21 years	0.429	
22 years	1.541		22 years	0.793	
Marital status			Marital status		
Unmarried	1		Unmarried	1	
Married	0.487		Married	1.421	
Per capita household income		***	Per capita household income		
<300 Euros	0.415		<300 Euros	0.228	
>300 Euros	1		>300 Euros	1	
Children in the household		***	Children in the household		
No child <18 years	1		No child <18 years	1	
1+ children <18 years	1.387		1+ children <18 years	0.514	
1999	1		1999	1	
2003	0.959	***	2003	1.217	
2005	0.702		2005	1.182	
Measures of model fit			*Measures of model fit*		
No. of observations	10,795		No. of observations	9,450	
Log-likelihood (0)	−1.507		Log-likelihood (0)	−137	
Log-likelihood	−1.452		Log-likelihood	−131	
LR [=LR chi2]	109		LR [=LR chi2]	11	
McFadden's R2	0.04		McFadden's R2	0.04	
df	13		df	13	

than that of males living in the reference state Saxony (OR = 2.582 in Model 18). In West Germany, males from Lower Saxony (OR = 2.220 in Model 19) and Schleswig-Holstein (OR = 1.998 in Model 19) are particularly prone to basic military service. Urbanity, in terms of a larger settlement size, has a consistently

Table 5.15 Results, basic military conscript (OR), Germany, males

Model 17

Germany, males	OR	p
Germany		
West	1	
East	1.934	***
Settlement size		
<20,000 residents	1	
>20,000 residents	0.875	**
Educational attainment		
Middle level or below	1	
Upper level	0.487	***
Age group		
19 years	1	
20 years	0.826	***
21 years	0.797	***
22 years	0.756	***
Marital status		
Unmarried	1	
Married	2.381	**

Model 18

East Germany, males	OR	p
Federal state		
Berlin	0.903	
Brandenburg	1.163	
Mecklenburg	2.582	***
Saxony	1	
Saxony-Anhalt	0.881	
Thuringia	1.034	
Settlement size		
<20,000 residents	1	
>20,000 residents	0.873	
Educational attainment		
Middle level or below	1	
Upper level	0.567	***
Age group		
19 years	1	
20 years	0.650	***
21 years	0.540	***
22 years	0.445	***
Marital status		
Unmarried	1	
Married	1.301	

Model 19

West Germany, males	OR	p
Federal state		
Baden-Württemberg	1	
Bavaria	1.519	***
Bremen	1.503	*
Hamburg	1.456	
Hesse	1.213	
Lower Saxony	2.220	***
North Rhine-Westph.	1.426	***
Rhineland-Pfalz	1.546	***
Saarland	0.833	
Schleswig-Holstein	1.998	***
Settlement size		
<20,000 residents	1	
>20,000 residents	0.887	*
Educational attainment		
Middle level or below	1	
Upper level	0.441	***
Age group		
19 years	1	
20 years	1.043	
21 years	1.028	
22 years	1.053	
Marital status		
Unmarried	1	
Married	2.709	**

(continued)

Table 5.15 (continued)

	Model 17			Model 18			Model 19		
	Germany, males	OR	p	East Germany, males	OR	p	West Germany, males	OR	p
Per capita household income									
<300 Euros		3.324	***		2.756	***		3.512	***
>300 Euros		1			1			1	
Children in the household									
No child <18 years		1			1			1	
1+children <18 years		1.052			0.994			1.121	
1998		1			1			1	
1999		1.268	***		1.437	**		1.206	
2000		1.014			0.977			1.047	
2001		1.052			1.180			0.995	
2002		0.890			0.789			0.938	
2003		0.877			0.810			0.903	
2004		0.817	*		1.032			0.705	**
2005		0.750	**		0.436	***		0.885	
Measures of model fit									
No. of observations		6,405			1,852			4,553	
Log-likelihood (0)		−4.350			−1.280			−2.998	
Log-likelihood		−3.932			−1.179			−2.680	
LR [=LR chi2]		836			201			637	
McFadden's R2		0.10			0.08			0.11	
df		16			20			24	

negative effect on the likelihood of being enlisted in the military as a basic service conscript (e.g. OR = 0.875 in Model 17).

The estimation results for the remaining factors of influence turn out to be different for basic military conscripts than for male regular or professional soldiers. Hence, there is a strong negative association between educational attainment and basic military training. In the model for all eligible males in Germany, a better educated individual is only 50 % as likely to be a conscript as a less educated individual with otherwise similar socio-demographic traits (OR = 0.487 in Model 17). At the same time, there is a negative association between age and basic military service. Males at 22 years are about 25 % less likely to undergo basic military training than males at 19 years, which, in all likelihood, stems from the circumstance that the majority of males completes basic military training right after school, i.e. around the age of 19 years (OR = 0.756 in Model 17). Meanwhile, marriage has a marked, positive influence on the likelihood of basic military training (OR = 2.381 in Model 17).

Another strong predictor of undergoing basic military service is a less affluent social background. This applies to East and West Germany in nearly equal measure. For example, East German males that originate from a family with less than 300 Euros of disposable income for each individual are nearly three times as likely to be a basic military conscript as their better-to-do peers (OR = 2.756 in Model 18). In West Germany, the association between low social class and serving as a conscript is even more pronounced (OR = 3.512 in Model 19). In view of the second family background variable, the presence of younger dependents in the household is largely uninfluential on the likelihood of basic military training. The chances of serving as a conscript are similar for those with younger siblings and those without siblings (e.g. OR = 0.994 in Model 18). Lastly, and in line with the initial prediction, there is a strong negative association of time and basic military training. However, the nature of the effect evolves somewhat differently in the two parts of Germany. In West Germany, the likelihood of basic military training declined to 88 % of the base level by 2005 (OR = 0.885 in Model 19). In East Germany, the pattern of change is more volatile. In 2001, males had a 20 % higher likelihood to serve as a conscript than males in the reference year 1998 (OR = 1.180 in Model 18). In 2005, those chances were about 55 % lower than the reference year (OR = 0.436 in Model 18).

With regard to the effect of health (see Table 5.16), it becomes obvious that, in contrast to the initially formulated prediction, males that are overweight or obese are, in fact, more likely to undergo basic military training than males of normal weight (OR = 1.211 in Model 20). However, this aggregated figure for the German male population masks some distinct differences between the effect of body composition on basic military training in East and West Germany. In the former, overweight or obese males are about 15 % *less* likely to serve as military conscripts (OR = 0.857 in Model 21), while in the latter, overweight or obese males are nearly 40 % *more* likely to undergo basic military training relative to their peers of normal weight (OR = 1.389 in Model 22). At the same time, the impact of smoking behavior on the likelihood of basic military training is consistent in all models, albeit in contrast to the initial expectation. Independent of his origin from East or West Germany, a male smoker is about 20 % less likely to undergo basic military training than a male non-smoker (e.g. OR = 0.793 in Model 21).

Table 5.16 Results, basic military conscript (OR), Germany, males, health

Model 20

Germany, males	OR	p
Germany		
West	1	
East	1.519	***
Settlement size		
<20,000 residents	1	
>20,000 residents	0.773	*
Educational attainment		
Middle level or below	1	
Upper level	0.612	***
Health		
Normal weight	1	
Overweight/Obese	1.211	
Smoking behavior		
Non-smoker	1	
Smoker	0.772	*

Model 21

East Germany, males	OR	p
Federal state		
Berlin	0.331	**
Brandenburg	0.789	
Mecklenburg	0.856	
Saxony	1	
Saxony-Anhalt	0.945	
Thuringia	0.463	
Settlement size		
<20,000 residents	1	
>20,000 residents	0.916	
Educational attainment		
Middle level or below	1	
Upper level	0.514	*
Health		
Normal weight	1	
Overweight/Obese	0.857	
Smoking behavior		
Non-smoker	1	
Smoker	0.793	

Model 22

West Germany, males	OR	p
Federal state		
Baden-Württemberg	1	
Bavaria	1.176	
Bremen	1.338	
Hamburg	1.538	
Hesse	0.921	
Lower Saxony	1.943	*
North Rhine-Westph.	0.951	
Rhineland-Pfalz	1.023	
Saarland	0.986	
Schleswig-Holstein	0.934	
Settlement size		
<20,000 residents	1	
>20,000 residents	0.733	
Educational attainment		
Middle level or below	1	
Upper level	0.665	**
Health		
Normal weight	1	
Overweight/Obese	1.389	
Smoking behavior		
Non-smoker	1	
Smoker	0.795	

	Model 1	Model 2	Model 3
Age group			
19 years	1	1	1
20 years	0.818	0.698	0.905
21 years	0.673 *	0.475 *	0.803 ***
22 years	0.772	0.592	0.870
Marital status			
Unmarried	1		1
Married	0.486		1.002 **
Per capita household income			
<300 Euros	1	1	3.065
>300 Euros	2.966 ***	3.545 ***	1
Children in the household			
No child <18 years	1	1	1
1+children <18 years	1.096	0.706	1.328
1999	1	1	1
2003	0.559	0.461	0.578
2005	0.601	0.506	0.648
Measures of model fit			
No. of observations	865	262	603
Log-likelihood (0)	−586	−181	−402
Log-likelihood	−538	−162	−366
LR [=LR chi2]	96	38	72
McFadden's R2	0.08	0.11	0.09
df	13	16	21

5.6 Discussion

The main objective of this analysis was to explore the socio-demographic characteristics that lead youth to be interested in a military career, serve as regular or professional soldiers, and undergo basic military training. The findings support several broad conclusions: Consistent with other studies (e.g. Bachman et al. 2000: 7), gender differences in the prediction of military propensity were not as large as those for actual military service. However, the correlation between propensity and actual service appeared to be weaker for females than for males. In this regard, two facts have to be noted: Firstly, towards the end of the 1990s, females only accounted for 1.2 % of all German soldiers (Biehl et al. 2007: 179). Secondly, the comparativeness of the data (i.e. survey periods) is not quite warranted. Data from the Microcensus includes several years prior to the opening of all military careers to females in 2001. Data on enlistment propensity was collected in 2006, when more females had probably considered the occupational option of a military career.

Local origin turned out to be a consistent and strong predictor of both military propensity and military service. Accordingly, males (and females) from East Germany were much more likely to consider military enlistment and serve as a professional soldier or basic military service conscript. In particular, individuals from economically weak federal states with comparatively high rates of youth unemployment were prone to military service. In East Germany, this particularly applied to males from Mecklenburg and Saxony-Anhalt and, in West Germany, to males from Bremen, Lower Saxony, and Schleswig-Holstein.[9] At the same time, apart from the partially positive association between military service and residence in the city states of Bremen or Hamburg, a higher level of urbanity was associated with a lower likelihood of military service.

Educational attainment was strongly linked to enlistment propensity and actual military service. However, unexpectedly, the effect of being better educated worked in different directions in East and West Germany, at least when military service as a regular or professional soldier was concerned. While in West Germany, educational attainment was generally negatively linked with enlistment propensity and military service, better educated males in East Germany and better educated females in Germany were more likely to serve in the military than their less educated peers. Hence, it appears that, at least in the case of East German males and females, the professional segment of the Bundeswehr, indeed, lives up to the maxim of the all-volunteer force, namely "to maintain high standards for its soldiers so as to ensure a professional and competent force" (Korb and Duggan 2007: 467).

---

[9] For East Germany, the reference state Saxony had an average youth unemployment rate of 19.5 % between 1998 and 2005. The respective values for Mecklenburg and Saxony-Anhalt were 20.6 and 21.5 %. For West Germany, the average youth unemployment rate ranged at 7.2 % between 1998 and 2005, while it amounted to 15.1 % in Bremen, 11.1 % in Lower Saxony, and 10.7 % in Schleswig-Holstein (INKAR 2007).

To a certain extent, this favorable connection between human capital endowment and military service also applied to health-related factors. In the case of males, being overweight or obese was generally linked to a lower likelihood of military service. Yet, for females, there was a strong positive association between overweight or obesity and membership in the Bundeswehr. However, in this connection, it should be noted that over the period of observation, the majority of females served in non-combat roles or combat-support functions (Biehl et al. 2007: 179), where, in all likelihood, health-related factors and physical exercise played a less significant role than for males that participated in combat-related functions at a relatively higher rate. At the same time, smoking status did not have a marked influence on the likelihood of military service.

Among family background factors, low per capita household income turned out to be a strong negative predictor of serving as a regular or professional soldier, while the presence of younger children in the household increased the likelihood of basic military service. The generally low likelihood of military service among poor youth seems to contradict the notion of military service as a path of upward social mobility. However, the present findings are consistent with previous research, which demonstrated that recruitment for the all-volunteer force did not attract the lowest or highest socio-economic groups. Applicants of the lowest social strata were disproportionately rejected for military service due to poor health, low cognitive ability or criminal records 24 (Segal and Segal 2004: 24). Meanwhile, the positive association between the presence of younger dependents in the household and military service seems to confirm the initial hypothesis that, provided that military entry requirements are fulfilled, limited familial resources work as a "pull factor" towards military service for youth to acquire skills and generate income.

In view of the predictors of basic military service training, the pattern of results suggests that the Bundeswehr does not attract the preferred high-quality recruits by means of conscription. Rather, there is a generally negative association between basic military training and the educational attainment of males, regardless of whether they originate from East or West Germany. In addition, overweight or obese males have a consistently higher likelihood to serve as basic military conscripts than males of normal weight.

The preceding analysis evokes two general conclusions in connection with the selectivity of military service and the efficacy of conscription as a recruitment tool for high-quality personnel: Firstly, the comparison of estimation results by region suggests that males from West Germany that choose *not to enter* the Bundeswehr as a regular or professional soldier are generally more likely to be higher educated, originate from an urban setting, have a better body composition, stem from a well-to-do family, and have fewer siblings than those entering the Bundeswehr. The same comparison for males from East Germany reveals that those who *enter* the Bundeswehr as regular or professional soldiers are more likely to be higher educated, possess a better body composition, and have more siblings than those not entering the Bundeswehr. Thus, it appears that the Bundeswehr "skims" the high-quality male population in Eastern Germany. Secondly, conscription as a recruitment tool for the preferred high-quality youth seems largely ineffective. The presented analyses consistently show that primarily males with less favorable human capital traits are drawn into the Bundeswehr via the channel of conscription. Those with higher human capital endowments are more likely to enter the Bundeswehr as regular or professional soldiers instead.

5.7 Model and Data Evaluation

There are various strengths to the present study, including the apparently first systematic review of socio-demographic determinants of military propensity and military service in Germany. However, there are also a number of issues that limit the generalizability of the findings. Given the limited publicly available data on military propensity and military service in Germany, these limitations cannot be remedied at this point, if need be specifically noted.

Most importantly, in the absence of data sources explicitly targeted at the exploration of factors associated with military propensity and military enlistment in Germany, the explanatory power of the analyses suffers from a lack in covariate information. To that effect, previous research unanimously demonstrated the influence of other family, demographic, and educational background factors. These include youth's learning proficiency as measured in the military entrance examination, ability to finance further education, (Hosek and Peterson 1986: 9), parental influence on youth's educational expectations (Kilburn and Klerman 2000: 10; Nieva et al. 1997: 24), parental education (e.g. Bachman et al. 2000: 16), and parents' military experience (e.g. Thomas 1984: 307). Other established individual-level determinants for joining the military are largely motivational and can be grouped in four categories: (1) "Institutional", including the desire to serve, patriotism, wish for adventure or challenge, and desire to be a soldier; (2) "future oriented", including the desire for a military career or a scholarship for higher education; (3) "occupational", including the ambition to support one's family, lack of better options or strive for job security; and (4) "pecuniary", including the desire for the enlistment bonus and a well-paid job (Woodruff et al. 2006: 360). In the external military environment, there are other factors that have been found to influence youth's civilian job opportunities and thus, their interest in the military as a potential employer. These include the unemployment rate, the civilian-military pay ratio, job security and educational opportunities. In the internal military environment, the state of recruiting resources, in terms of the number of recruiters, advertising, educational benefits, and cash bonuses were established as major influential factors of military recruiting (Orvis and Asch 2001: 8). The inclusion of such variables in the collection of military-related information could increase the informative value of future studies about military propensity and military enlistment in Germany.

The validity of the presented propensity measures is somewhat restricted by the lack of information about the eventual enlistment decision of respondents (cf. Orvis et al. 1996: 4). However, previous research demonstrated that stated propensity for military service was a reliable predictor of future enlistment behavior (Woodruff et al. 2006: 354; Bicksler and Nolan 2006: 6). The consistent pattern of estimation results throughout the analyses seems to confirm that socio-demographic factors that are correlated with military propensity also influence the likelihood of actual military service.

There is a mixing of supply and demand effects given the nature of military enlistment as a "two-party decision" (Bachman et al. 2000: 6). Hence, recruiting outcomes are not only the result of supply factors, such as youth vocational prefer- ences, but are also influenced by the quantitative and qualitative manpower demand specified by the military (Dertouzos 1985: 1). Enlistment in the Bundeswehr does not constitute a unilateral decision but entails a positive agreement by the military recruiting service and the individual applicant. On this account, the issue of causa- tion remains unclear, in particular whether the higher odds of military service among youth with certain socio-demographic characteristics result from inherent vocational aspirations or whether the military, in fact, demands a higher share of that particular human capital type. Therefore, ascribing soldiers' human capital endowments, or changes therein over time, solely to shifts in supply could lead to premature conclusions (Dertouzos 1985: 6).

It should also be mentioned that the Microcensus uses a rotating interview scheme that surveys the inhabitants of a given household up to four times. As a result, there may be a duplication of records in the sample since the sets of respon- dents overlap in the years under observation. Since households or individuals can- not be identified across survey waves, it is not possible to account for this limitation. However, as suggested by Geisler and Kreyenfeld (2009: 18), the robustness of the results was verified by running separate analyses for the survey years 1999 and 2005, which do not include repeated records for the same individual. The respective estimates were comparable to the estimation results reported above.

Moreover, the specification of the binary outcome variable that identifies individuals as being enlisted in the military as a regular or professional soldier vs. pur- suing other civilian employment suffers from two major, yet non-detachable, limita- tions. On the one hand, the occupational category of being a regular or professional soldier also includes members of the Federal Police and riot police. On the other, it does not differentiate between military and civilian personnel in the Bundeswehr. Both misclassifications may lead to an overestimation of the probability of youth to select the soldierly profession for a career. As to the inclusion of the Federal Police and riot police in the sample, bias may arise from the divergent range of duties, both in terms of content and location, as well as the unequal risks and hardships that arise from service. Therefore, the probability of an individual with certain socio- demographic attributes to serve in the military may vary from that of being a police officer. In view of the aggregation of military and civilian personnel, the issue of overestimation seems particularly relevant for females whose relative share in the combat segment is comparatively small.

In the absence of data about military enlistment during the first 1 or 2 years after graduation from school, as taken advantage of prior U.S.-based research (e.g. Bachman et al. 2000), current occupational status of *being enlisted* as a soldier or a basic conscript are considered as a proxy for determining socio-demographic factors associated with initial military *enlistment*. Two strategies have been considered to minimize the inherent bias in this approach. On the one hand, the age range under consideration is defined rather narrow to only include people between 19 and 22 years. It is assumed that the likelihood of military enlistment, both as a recruit and professional soldier, is highest then, and chances are that a substantial share of the respondents still lives in the parental home. On the other hand, it has been considered to only include respondents in the sample that have not changed their place of residence during the past 12 months, assuming that, in all likelihood, these people still live with their parents and therefore provide information of their socio-demographic background that is less fraught with uncertainty. However, as response to the respective question was voluntary, item non-response was high and would have reduced the sample significantly without guaranteeing that information about respondents' socio-demographic background was more accurate.

The nature of the Microcensus data makes a straightforward identification of family background and regional origin difficult. The hypotheses that people that originate from less affluent social backgrounds, families with a larger number of children, economically weak areas or less urban areas are more prone to join the military are difficult to examine by means of the Microcensus. In contrast to the German Socioeconomic Panel, it does not provide ready-to-use family background variables. Explanatory variables used as proxies for respondents' socio-demographic background include the per capita household income, the number of children below the age of 18 in the household, and in terms of regional origin, the place of residence (East vs. West Germany or federal state) and the settlement size. However, regression estimates may be biased due to the fact that the Microcensus only inquires *current* per capita household income, the *current* number of children below 18 years in the household, and details about the *current* place of residence. Potential bias in the estimates may stem from the lack of knowledge about whether the respondent still lives with his/her parents and provides details about his/her socio-demographic background, or whether he/she already moved out of the parental home and thus specifies his/her independently earned living conditions. In the latter case, there would be a confounding of family background, personal living circumstances, and even occupational characteristics.

For instance, with regard to per capita household income, it remains unclear whether respondents provide details on the income situation of their family of origin or whether they specify their own income, with recruits and professional soldiers, in fact, specifying income realized during military service. In this context, it also needs to be mentioned that per capita household income is inasmuch inferior to the commonly used indicator of disposable household income per individual as it does not adjust for differences in household size and remains inexplicit about the inclusion of total market income (gross earnings plus gross capital and self-employment incomes), transfers from public sources and deductions from income taxes and social security contributions, which all flow into the calculation of the alternative equivalent disposable household income per individual (Förster and Pearson 2002: 10). In view of the nature of the Microcensus data, the calculation of this preferred measure of per capita household income seems problematic given the classification of household income into income classes rather than inquiring exact values, which would be required for the calculation of the equivalent disposable household income per individual (Strengmann-Kuhn 1999: 379). Moreover, it remains unknown whether the children below the age of 18 that live in the household are, in fact, the siblings of the respondent or in any other way related. In addition, the stated place of residence (East vs. West Germany or federal state) and the size of the settlement do not give absolutely certain information about the regional origin of the respondent. In the worst case, a respondent, such as a military recruit or professional soldier, moved to a small settlement in East Germany due to his profession, thereby contributing to an overestimation of both explanatory variables.

5.8 Conclusion

The present analysis demonstrated a consistent pattern of predictive associations between socio-demographic background variables and military propensity or military service in Germany. However, inferences about the viability of military service in the future are hampered by two foreseeable developments: (1) Statistical relations between socio-demographic traits and military service may change in the future as potential enlistees develop a real expectation of combat. (2) Demographic change, particularly in East Germany, will reduce the size and alter the composition of the youth population available for military recruitment. First and foremost, this will improve youth's general labor market outlook and increase the competition for skilled personnel. However, there are also other implications of demographic change that become evident in a review of the main determinants of military propensity and military service.

The analysis proved a consistently positive predictive association between military service and residence in Mecklenburg-Western Pomerania and Saxony-Anhalt in the case of East Germany, and Lower Saxony and Schleswig-Holstein where West Germany is concerned. While the decline in youth will be steep in all five East German states until 2030, the change in Mecklenburg-Western Pomerania and Saxony-Anhalt will be particularly dramatic. Similarly, projections of the absolute number of youth in Lower Saxony and Schleswig-Holstein are gloomy relative to most other federal states in West Germany. On the other hand, several city states and federal states feature a reasonably favorable demographic outlook. These include Hamburg, Bremen, Baden-Württemberg, Bavaria, and Hesse (Federal Statistical Office 2006a). Unfortunately, with the exception of Bremen, residence in these regions was associated with a consistently low likelihood to serve in the military.

With regard to the influence of urbanity, youth from smaller towns were consistently most likely to consider a military career and serve in the Bundeswehr. Yet, peripheral and old-industrialized regions will be most severely affected by demographic change. They are headed for a fast decline in youth and a rapid increase in the elderly population share along with a progressive decline in the total population. In the meantime, metropolitan areas, especially in West Germany, including

Hamburg, Frankfurt, Munich and Stuttgart, are projected to continue to grow (Höhn et al. 2007: 29).

The influence of educational attainment on military service was divided: In East Germany, better educated individuals were more likely to serve in the military, while in West Germany, educational attainment was inversely related to military service. In this regard, Easterlin (1978: 401) argued that with a smaller youth cohort size and the retirement of older workers, the educational and occupational opportunity structure of young males would improve considerably, solely on the basis of demographic change. On the assumption that civilian institutions will launch attractive recruitment incentives, the opportunity costs of military enlistment will increase as the youth population declines. In all likelihood, this will especially apply to young and highly-skilled youth in East Germany.

In terms of educational trends, Seitz and Kempkes (2007: 395) expect a continuously high demand for tertiary education. They estimate that the enrolment rate will remain stable at the current level of 35 % until the year 2030. Other concerns have been raised about the impact of the shrinking working-age population on the average wage level, and hence, on the competitiveness of military pay. Provided that demographic change will lead to an increase in average wage rates and a decline in youth unemployment below current levels, civilian employment could become even more attractive to high-quality youth compared to a military career (Dertouzos 1985: 1; Wilson et al. 1988: 2).

The overlap of the military target population with that of other major civilian institutions (Sandell 2006: 85) and the tendency of those with an above-average learning proficiency to prefer university enrolment or a civilian profession to a military career (Bachman et al. 2000: 4) will likely constitute the most significant obstacles to the recruitment of high-quality youth in the future. In light of the presented estimation results and official population projections, the anticipated improvements in the educational and occupational opportunity structure of youth will significantly reduce the quantitative and qualitative recruitment potential, most notably in East Germany. Meanwhile, the demography-induced budgetary tradeoffs will make it progressively more difficult to offer attractive recruitment incentives and substitute capital for labor to compensate for potential manpower shortfalls (Jackson and Howe 2008: 90).

Chapter 6
The Military Recruitment Target Population

The projections are based on an extrapolative method and a modified cohort-component approach. All human capital information are reported for the ages 19–22 years and specifically for males *and* females, further separated by German and non-German nationality. The underlying rationale is that, while the Bundeswehr still focuses on recruiting from the male population, the foreseeable reduction in this core target group may result in a more active recruiting effort among other subgroups in the population. Given the particular importance of human factors for military effectiveness, a closer examination of the level and distribution of human capital endowments within these subpopulations seems of particular importance. This inclusion of qualitative information is essential for a realistic estimation of the future military recruitment potential. Other projections, solely based on quantitative information, may neglect important social trends that are detrimental to military recruitment and overestimate the true recruitment potential.

6.1 Quantity and Quality in Military Recruitment

While demographic change may constrain the feasibility of quantitative recruitment objectives in the future, any discussion of current and future military manpower supply remains incomplete when limited only to changes in quantity without considering the human capital endowments of the military recruitment pool and changes therein as a result of demographic shifts (cf. Boehmer et al. 2003: 41). Previous research suggests that military recruiting in contemporary industrialized societies will increasingly be complicated by trends in several key qualification areas including youth's educational attainment, aptitude levels, as measured by the military entrance examination, obesity, medical conditions (e.g. asthma, and diabetes), physical performance, moral conditions (i.e. prior criminal records), drug and alcohol misuse, and responsibility for dependents

W. Apt, *Germany's New Security Demographics: Military Recruitment*
in the Era of Population Aging, Demographic Research Monographs,
DOI 10.1007/978-94-007-6964-9_6, © Springer Science+Business Media Dordrecht 2014

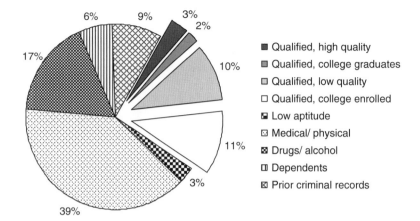

6%   9%   3%
                 2%

17%

                        10%

                                ■ Qualified, high quality
                                ■ Qualified, college graduates
                                ▢ Qualified, low quality
                                ▢ Qualified, college enrolled
                                ▣ Low aptitude
                                ▣ Medical/ physical
                        11%   ▨ Drugs/ alcohol
                                ▥ Dependents
                        3%   ▨ Prior criminal records

39%

Fig. 6.1 Eligibility for military service (Source: Bicksler/Nolan (2006))

(Boehmer et al. 2003: 71).[1] For the United States, it has been estimated that 7 out of 10 youth are currently ineligible for military service (see Fig. 6.1). While the main reduction in the military recruitment potential stems from youth' failure to meet the minimal entry requirements, others, though qualified for military service, prefer higher education at a civilian institution. Thus, the share of youth willing to enlist is even smaller than those eligible for military service. Ultimately, it is estimated that only 15 % of the total U.S. youth population belong to the available recruitment pool, with only one third of them being considered high quality (Bicksler/ Nolan 2006: 4). According to the National Research Council, the rates of various disqualifying physical and medical conditions or moral behaviors may further increase in the next two decades. In that, changes in drug use, obesity, and asthma will increase the share of the youth population ineligible for military service (NRC 2003: 259).

6.2   Trends in Young Adult Demography

The absolute number of young males and females in the typical target age group for recruitment to the Bundeswehr is projected to decline significantly until 2030. Figures 6.2 and 6.3 report this substantive reduction in the male and female population for single age groups between 19 and 22 years (left axis) and the aggregate age group (right axis).

---

[1] Boehmer et al. (2003: iv) estimated that 45 % of the U.S. youth population are ineligible for military service due to physical or medical condition. Thereof, 26 % have been diagnosed with asthma, attention deficit or other mental disorder, irregular blood pressure or diabetes. An additional 21 % of youth fail to meet the weight standards of the U.S. military, while another 15 % are restricted in their physical performance.

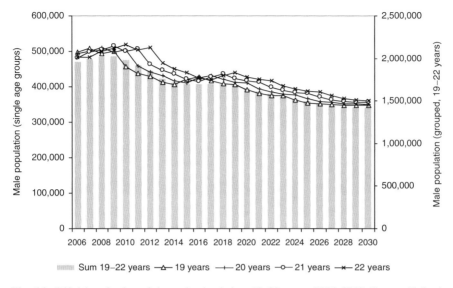

Fig. 6.2 Official projection of the male population, 19–22 years, 2006–2030 (Source: Federal Statistical Office (2006a), own estimations)

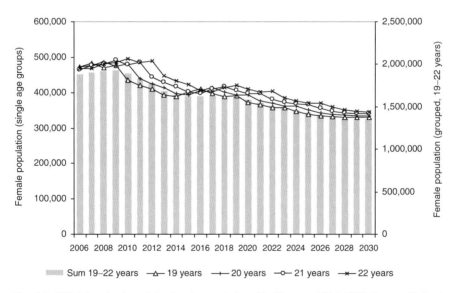

Fig. 6.3 Official projection of the female population, 19–22 years, 2006–2030 (Source: Federal Statistical Office 2006a, own estimations)

Despite a universal decline in the youth population, there will be some distinct regional differences in the development and magnitude of the demographic shift ahead. Figures 6.4 and 6.5 show the projected relative change in the male and female

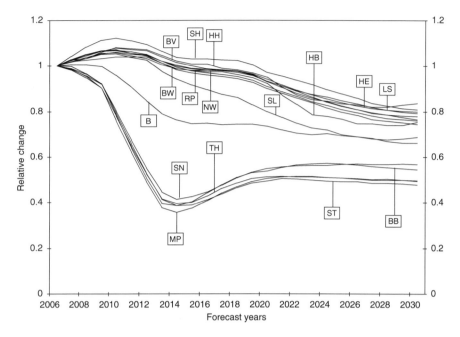

Fig. 6.4 Relative change in males, federal state, 19–22 years, 2006–2030 (Source: 11th coordinated population projection, regional variant (Federal Statistical Office 2006a), own estimations. *B* Berlin, *BB* Brandenburg, *BW* Baden-Württemberg, *BV* Bavaria, *HB* Hanseatic City of Bremen, *HE* Hesse, *HH* Hanseatic City of Hamburg, *MP* Mecklenburg-Western Pomerania, *LS* Lower Saxony, *NW* North Rhine-Westphalia, *RP* Rhineland-Pfalz, *SH* Schleswig-Holstein, *SN* Saxony, *SL* Saarland, *ST* Saxony-Anhalt, *TH* Thuringia)

population between the ages 19–22 across federal states. The most prominent feature consists in the demographic divide between East and West Germany. By 2030, the youth population of males (Fig. 6.4) and females (Fig. 6.5) in all East German states will have declined to about 50–60 % of the 2006 level. Over the same time period, the decline of the West German youth population will be less severe due to more favorable trends in fertility and migration in the past. By 2030, the size of the West German youth population across states will range between 65 and 85 % of the 2006 level.

The state-specific examination reveals a distinct correlation between regional demographic trends and economic development. While in East Germany, the economically sound state of Saxony is projected to undergo a somewhat less severe decline in youth, it is also the high-performance states or metropolitan areas in West Germany that feature comparatively favorable medium-term projections. This applies to Hamburg, Bavaria, Baden-Württemberg and Hesse.

While, on this account, the relatively favorable demographic development projected for Schleswig-Holstein and Lower Saxony is somewhat unexpected, it may, in fact, result from low out-migration. This is consistent with Höhn et al.

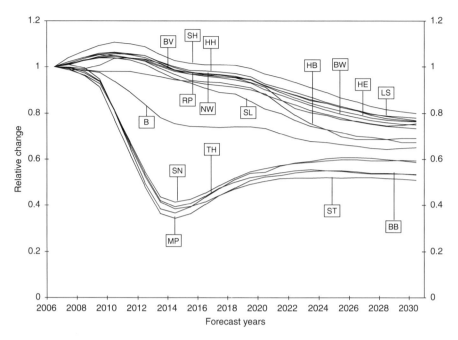

Fig. 6.5 Relative change in females, federal state, 19–22 years, 2006–2030 (Source: 11th coordinated population projection, regional variant (Federal Statistical Office 2006a), own estimations. *B* Berlin, *BB* Brandenburg, *BW* Baden-Württemberg, *BV* Bavaria, *HB* Hanseatic City of Bremen, *HE* Hesse, *HH* Hanseatic City of Hamburg, *MP* Mecklenburg-Western Pomerania, *LS* Lower Saxony, *NW* North Rhine-Westphalia, *RP* Rhineland-Pfalz, *SH* Schleswig-Holstein, *SN* Saxony, *SL* Saarland, *ST* Saxony-Anhalt, *TH* Thuringia)

(2007: 22) that emphasize internal migration as a major driver of population dynamics on lower levels of regional aggregation.

An additional demographic factor relevant to military recruitment is the proportion of national and non-national persons in the target population. A large influx of migrants since the early 1990s accounted for today's ethnocultural diversity in Germany. Tables 6.1 and 6.2 report current proportions of youth with German citizenship and foreign-born youth with EU- or non-EU citizenship. Evidence suggests that there are some distinct differences between federal states in the proportion of youth with German and foreign citizenship. In particular, federal states in West Germany display a more heterogeneous youth population than those in East Germany. Most notably, these include Baden-Württemberg, Berlin, Hamburg, Hesse, and North-Rhine Westphalia. Also noteworthy is the high share of persons with non-EU citizenship within these federal states. As a result, the average share of males with German citizenship is less than 87 % in West Germany, while it amounts to about 97 % in East Germany (Table 6.2). Similar results are reported for females (Table 6.3).

Table 6.1 Nationality by region, males 19–22 years

Federal state	Percentage of sample			Number of cases
	German	EU-Foreigner	Non-EU-Foreigner	
Baden-Württemberg	84.1	4.6	11.4	1,651,836
Bavaria	89.1	2.5	8.4	1,954,446
Berlin	85.2	1.3	13.5	603,234
Brandenburg	99.1	0.2	0.7	505,764
Bremen	89.3	1.2	9.5	150,282
Hamburg	85.2	2.6	12.2	501,847
Hesse	85.1	3.0	12.0	826,548
Lower Saxony	99.2	0.0	0.8	505,393
Mecklenburg	90.9	1.9	7.3	1,131,920
North Rhine-Westphalia	84.3	2.9	12.8	2,391,733
Rhineland-Pfalz	88.5	3.0	8.6	734,327
Saarland	89.6	3.2	7.2	494,507
Saxony	99.0	0.1	0.9	923,244
Saxony-Anhalt	99.0	0.1	1.0	531,617
Schleswig-Holstein	89.5	2.3	8.2	598,843
Thuringia	99.5	0.1	0.4	545,781
Germany	89.5	2.3	8.3	14,051,322
West	86.9	2.9	10.2	10,436,289
East	96.8	0.3	2.9	3,615,032

Source: Scientific use files of the German Microcensus 1998–2005, own estimations
Notes: Weighted to national level using sampling weights provided with the German Statistical
Office. Pearson chi-squared test: $p = 0.000$

6.3   Trends in Educational Attainment

The strict requirements of educational certification for entry into the Bundeswehr
necessitate careful consideration of the trends in educational attainment and their
interaction with military recruiting.

Since the 1970s, there has been noticeable trend towards higher education and a
tertiary knowledge society in Germany (Bonin et al. 2007). To that effect, the distribution of school graduation rates markedly changed: While graduation from lower
level education became less important, there has been a general trend towards graduation from medium level and higher level education (see Table 6.3).

Corresponding to this shift in school graduation rates, university enrollment has
been on the rise since the mid-1990s (see Fig. 6.6). Albeit a slowdown since 2004,
one third of high school graduates choose to enroll in a tertiary program at a university each year. Compared to the OECD average, tertiary enrollment rates in Germany
are still relatively low and will likely rise in the future.

This longer-term increase in university enrollment negatively affects military
recruitment since it does not only reduce the number of youth potentially interested
in pursuing a military career but also attracts a disproportionate share of the

Table 6.2 Nationality by region, females 19–22 years

	Percentage of sample			
Federal state	German	EU-Foreigner	Non-EU-Foreigner	Number of cases
Baden-Württemberg	83.5	4.8	11.6	1,598,364
Bavaria	88.1	2.9	9.0	1,920,043
Berlin	86.2	1.9	11.9	608,122
Brandenburg	98.8	0.3	0.9	402,226
Bremen	87.3	1.3	11.4	155,767
Hamburg	87.1	3.0	9.9	501,873
Hesse	83.1	3.5	13.4	828,756
Lower Saxony	98.4	0.2	1.4	439,354
Mecklenburg	90.5	1.5	8.0	1,077,827
North Rhine-Westphalia	83.6	2.8	13.6	2,324,637
Rhineland-Pfalz	89.6	2.6	7.8	690,058
Saarland	87.7	3.9	8.3	476,627
Saxony	99.3	0.1	0.6	814,800
Saxony-Anhalt	98.0	0.1	1.9	439,670
Schleswig-Holstein	88.5	3.3	8.2	551,094
Thuringia	99.4	0.1	0.5	480,099
Germany	88.7	2.5	8.8	13,309,315
West	86.3	3.1	10.6	10,125,045
East	96.4	0.5	3.1	3,184,270

Source: Scientific use files of the German Microcensus 1998–2005, own estimations
Notes: Weighted to national level using sampling weights provided with the German Statistical Office. Pearson chi-squared test: p=0.000

Table 6.3 School graduation rates, 1996 vs. 2006

School degree	Age group (years)	1996 %	2006 %	Δ %
Lower level (*Hauptschulabschluss*)	15–17	30.6	28.5	−2.1
Middle level (*Realschulabschluss*)	16–18	46.4	49.6	3.2
Higher level (*Fachhochschulreife*)	18–21	8.5	13.6	5.1
Higher level (*Abitur*)	18–21	28.0	30.0	2.0

Source: National Report on Education (2008: 60)

high-quality segment of the youth population that is generally preferred by the military (Bicksler and Nolan 2006: 6). Hence, youth lost to tertiary education at a civilian institution tend to have higher aptitudes than those who choose another avenue of professional development. The extent of this reduction is illustrated by a study of Warner et al. (2001: 29) that estimated that about one third of the decline in military propensity among White males in the United States between 1985 and 1997 could be attributed to an increase in university enrollment. Against this background,

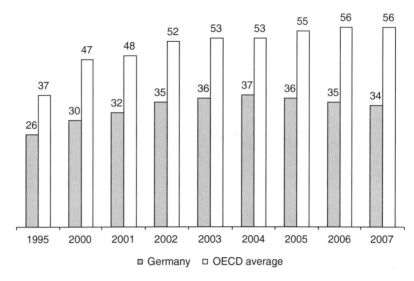

Fig. 6.6 Entry rates at tertiary level, 1995–2007 (Source: OECD (2009: 60))

the National Research Council (2003: 256) concluded: "The dramatic increase in college enrollment is arguably the single most significant factor affecting the environment in which military recruiting takes place."

The trend towards higher education could be interpreted as an indication of an improved level of competence among youth. However, this seems to be a foregone conclusion in view of (1) students' limited improvements in the areas of reading, mathematics and science; (2) a growing disparity between high-performance and low-performance students; (3) a lacking integration of students with migration background; and (4) consistently high school-dropout rates (cf. National Report on Education 2008). Thus, along with a trend towards higher education, there appears to be the opposing trend of an educational stagnation (e.g. Reinberg and Hummel 2003: 4; Hradil 2004: 141).

In 2000, the performance of 15-year-old German students in the Program for International Student Assessment (PISA) for the areas of reading competence, mathematics and science was below the average student performance in OECD countries (National Report on Education 2008: 82). At the same time, a high ratio of so-called "problem students" was identified: 23 % of 15-year-old students were incapable to comprehend the central theme of a short text at the end of their compulsory general education (Artelt et al. 2001b: 99). About 17 % of students only possessed rudimentary mathematical skills comparable to the level of an elementary school student, making them incapable to solve arithmetic problems typical for applicant apprentices (Klieme et al. 2001: 170). In the domain of science, the 2000 PISA study documented major differences in students' capability and considerable deficits in basic science education. About 26 % of the tested 15-year-old students scored the lowest proficiency level I; another 26 % were proficient on level II (Artelt et al. 2001a: 29). These figures cause concern seeing that: "At Level 2, students start to demonstrate the science competencies that will enable them to actively participate in life situations related to science and technology" (OECD 2008: 106).

Between 2000 and 2006, average achievement levels of 15-year-old students on the PISA tests increased for mathematics and the sciences, yet stagnated in the area of reading competence and still featured significant variation in students' cognitive ability (cf. National Report on Education 2008). Also noteworthy is the strong association between social background and student performance in Germany. For example, students with a migration background seem to be largely disadvantaged in the school system and their educational achievement. In particular, second-generation migrants show significant performance deficits, which even deepened between 2000 and 2006 (National Report on Education 2008: 85).

Equally worrisome is the unsuccessful school-to-career transition among a significant number of youth. Over many years now, the proportion of individuals leaving school without a certificate at the age of 15–17 years has ranged at a steady level of about 8 %, corresponding to about 76,000 students per year. School dropout rates vary by population group: more males than females and twice as much students with a migration background leave school without final qualification (National Report on Education 2008: 90).

It becomes obvious that the target group of the Bundeswehr is split by an educational divide that makes recruitment more difficult in quantitative and qualitative terms. While a large portion of talented youth pursues higher education and is therefore unavailable for military recruitment, another significant share of youth does not fulfill the cognitive requirements of a professional career in the Bundeswehr at the end of their compulsory general education.

When the core target population between 19 and 22 years is concerned, there are significant differences in the distribution of education by region and nationality. Table 6.4 reports the educational attainment level of German males and indicates strong distinctions between East and West Germany, in particular for the levels of low and middle education. In East Germany, the majority of males (53.4 %) obtains a middle level education. Lower level schooling appears less common. At the same time, the share of East German males with a university-entrance certificate is smaller compared to West Germany. There, the distribution of educational attainment levels is well-balanced with about one third of males completing a low, middle or high level education. In Baden-Württemberg, Hesse, North Rhine-Westphalia and Saarland, the educational distribution appears particularly favorable, and the share of German males with a higher level of schooling ranges above 35 %.

In comparison, as shown in Table 6.5, males with foreign citizenship more often obtain lower level education than their German peers (54.8 % vs. 28.7 %). Particularly in Baden-Württemberg and Bavaria, the distribution of educational attainment among foreign males leans towards the lower level (59.6 and 64.3 %). Conversely, in East Germany, the share of foreign males with an upper level education is unusually high.

Table 6.4 Educational attainment by region, Males 19–22 years, German

| Federal state | German males | | | |
	Lower level (Hauptschule)	Middle level (Realschule)	Upper level (Gymnasium)	Number of cases
Baden-Württemberg	31.2	33.8	35.0	1,338,681
Bavaria	43.2	31.9	24.9	1,674,119
Berlin	22.4	42.1	35.5	494,859
Brandenburg	14.1	59.4	26.5	486,365
Bremen	20.5	48.5	31.0	127,274
Hamburg	34.7	32.0	33.3	412,626
Hesse	26.1	37.8	36.1	681,933
Lower Saxony	22.2	53.7	24.1	489,219
Mecklenburg	29.5	41.8	28.7	990,409
North Rhine-Westphalia	29.4	33.2	37.4	1,932,498
Rhineland-Pfalz	34.3	33.2	32.4	629,800
Saarland	30.9	30.5	38.6	423,362
Saxony	16.7	54.0	29.3	897,442
Saxony-Anhalt	15.7	59.7	24.6	513,701
Schleswig-Holstein	30.9	36.8	32.3	512,390
Thuringia	21.3	51.2	27.5	530,349
Germany	28.73	39.91	31.36	12,135,026
West	32.73	34.62	32.65	8,723,092
East	18.5	53.44	28.07	3,411,934

Source: Scientific use files of the German Microcensus 1998–2005, own estimations
Notes: The educational attainment category "missing", which contains the sub-categories "no information" and "no degree", is omitted. Weighted to national level using sampling weights provided with the German Statistical Office. Pearson chi-squared test: $p = 0.000$

Over the period from 1998 to 2005, the distribution of education among males with German citizenship improved. The share of males with the highest level of school education increased, and the share of males with the lowest level of schooling declined accordingly (see Fig. 6.7). At the same time, males with a foreign citizenship substantially improved their educational achievement. A decreasing share of foreign males completed a lower level of schooling, and an increasing share attended programs in middle level and upper level institutions of education (see Fig. 6.8).

Among females, the educational differences by region and nationality are similar. Native females from East Germany graduate less often from lower level schools than females from West Germany (10.3 % vs. 20.8 %). The share of West German females with a lower level school degree is particularly large in Bavaria (29.3 %) and Hamburg (23.8 %). In East Germany, the majority of females obtains a middle level school degree (47.8 %). In West Germany, graduation rates from middle and higher level institutions of education both range around 40 % among females. In Berlin, North-Rhine-Westphalia, and Saarland, the share of more highly educated females with German citizenship ranges above 45 % (see Table 6.6).

Table 6.5   Educational attainment by region, Males 19–22 years, foreign

	Foreign males			
	Lower level (Hauptschule)	Middle level (Realschule)	Upper level (Gymnasium)	Number of cases
Baden-Württemberg	59.6	19.4	21.0	242,691
Bavaria	64.3	15.2	20.5	195,274
Berlin	37.0	25.9	37.1	77,441
Brandenburg	41.4	33.7	24.9	3,107
Bremen	50.6	19.0	30.5	14,719
Hamburg	49.6	27.0	23.4	64,807
Hesse	49.7	28.0	22.3	115,520
Lower Saxony	0.0	70.3	29.7	3,033
Mecklenburg	52.3	28.9	18.8	86,707
North Rhine-Westphalia	54.6	21.5	23.8	335,442
Rhineland-Pfalz	57.5	22.7	19.9	78,171
Saarland	48.6	26.3	25.2	42,866
Saxony	15.6	35.8	48.6	8,345
Saxony-Anhalt	30.6	46.4	23.1	4,283
Schleswig-Holstein	59.7	24.7	15.6	55,358
Thuringia	46.4	13.5	40.1	2,430
Germany	54.8	22.4	22.9	1,330,195
West	56.4	21.9	21.7	1,231,554
East	34.1	29.0	36.9	98,641

Source: Scientific use files of the German Microcensus 1998–2005, own estimations
Notes: The educational attainment category "missing", which contains the sub-categories "no information" and "no degree", is omitted. "Foreigners" include EU-Foreigners and Non-EU-Foreigners. Weighted to national level using sampling weights provided with the German Statistical Office. Pearson chi-squared test: $p = 0.000$

The share of females with foreign citizenship that obtain a lower level of education is more than twice as high than among German females, regardless of whether foreign females in East or West Germany are concerned (see Table 6.7). The proportion of foreign females with an upper level school degree is generally lower than among German females. However, those in East Germany, just as their male counterparts, graduate at unexpectedly high rates (38.2 %) from upper level institutions. This mainly stems from the favorable educational distribution in Saxony, Saxony-Anhalt, and Thuringia although smaller sample sizes in East Germany limit the generalizability of the results.

Over time, the educational distribution among German females shifted towards higher educational attainment. The share of females with a higher level of schooling increased from 39.6 % in 1998 to 41.7 % in 2005 (see Fig. 6.9). Among foreign females, the gains were even more pronounced (see Fig. 6.10). Between 1998 and 2005, there was a marked decline in the share of foreign females with a lower level degree (50.8 % vs. 36.6 %) and an almost identical increase in the share of graduates from upper level schools (21.8 % vs. 35.6 %).

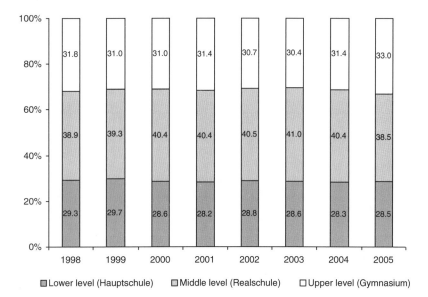

Fig. 6.7 Educational attainment, 1998–2005, Males 19–22 years, German (Source: Scientific use files of the German Microcensus 1998–2005, own estimations. Notes: Weighted to national level using sampling weights provided with the German Statistical Office. Pearson chi-squared test: p = 0.001. Educational attainment categories "No information" and "No degree" were omitted due to small number of cases and for consistency. Sample size: 1998 = 1,444,111; 1999 = 1,467,645; 2000 = 1,502,824; 2001 = 1,544,805; 2002 = 1,524,448; 2003 = 1,492,706; 2004 = 1,498,504; 2005 = 1,659,983)

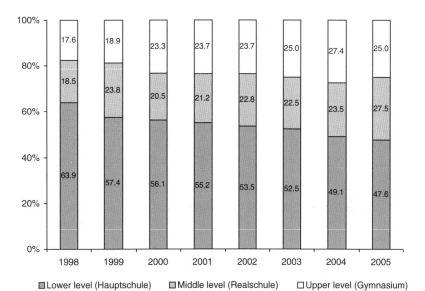

Fig. 6.8 Educational attainment, 1998–2005, Males 19–22 years, Foreign (Source: Scientific use files of the German Microcensus 1998–2005, own estimations. Notes: Weighted to national level using sampling weights provided with the German Statistical Office. Pearson chi-squared test: p = 0.000. Educational attainment categories "No information" and "No degree" were omitted due to small number of cases and for consistency. Sample size: 1998 = 182,912; 1999 = 175,426; 2000 = 168,176; 2001 = 178,307; 2002 = 175,834; 2003 = 162,526; 2004 = 140,057; 2005 = 146,956)

Table 6.6 Educational attainment by region, Females 19–22 years, German

| Federal state | German females | | | |
	Lower level (Hauptschule)	Middle level (Realschule)	Upper level (Gymnasium)	Number of cases
Baden-Württemberg	18.6	41.7	39.7	1,292,854
Bavaria	29.3	39.8	30.9	1,634,572
Berlin	13.4	39.7	46.9	509,151
Brandenburg	8.5	49.3	42.3	388,762
Bremen	15.9	39.4	44.7	129,564
Hamburg	23.8	38.2	38.1	423,081
Hesse	14.6	39.0	46.5	670,596
Lower Saxony	11.9	50.9	37.2	423,618
Mecklenburg	17.2	46.9	35.9	942,585
North Rhine-Westphalia	19.4	33.9	46.7	1,866,356
Rhineland-Pfalz	21.3	38.0	40.8	599,872
Saarland	19.1	33.5	47.5	405,617
Saxony	9.1	48.1	42.8	800,589
Saxony-Anhalt	9.1	53.1	37.8	421,778
Schleswig-Holstein	19.6	43.3	37.2	474,211
Thuringia	10.0	47.4	42.7	465,959
Germany	18.1	41.5	40.5	11,449,166
West	20.8	39.2	40.0	8,439,308
East	10.3	47.8	41.9	3,009,857

Source: Scientific use files of the German Microcensus 1998–2005, own estimations
Notes: The educational attainment category "missing", which contains the sub-categories "no information" and "no degree", is omitted. Weighted to national level using sampling weights provided with the German Statistical Office. Pearson chi-squared test: $p=0.000$

Based on the current educational distribution and various assumptions about transition rates in the future, the Standing Conference of the Ministers of Cultural Affairs of the Federal States of Germany provides estimates of the number of graduates in East and West Germany between 2000 and 2015, separated by school type. For the purpose of the present analysis, the official graduation rates were carried forward until 2030 by means of a logarithmic trend extrapolation. The projected estimates are reported in Figs. 6.11, 6.12 and 6.13.

Throughout Germany, the absolute number of graduates from lower and middle level institutions of education is projected to decline, just as the number of youth leaving school without a certificate (see Fig. 6.11). At the same time, the number of graduates with upper level education is projected to increase until 2030.

Turning to the projected number of graduates in East and West Germany, strong distinctions emerge. In West Germany, the absolute number of graduates with middle and higher level education is expected to increase markedly, while the number of students with lower level education and without final educational certification is projected to decline (see Fig. 6.12). Thus, the foreseeable decline in the German youth population is not entirely reflected in the West German educational system, which is projected to experience a considerable increase in the absolute number of students in higher education.

Table 6.7 Educational attainment by region, Females 19–22 years, foreign

	Foreign females			
	Lower level (Hauptschule)	Middle level (Realschule)	Upper level (Gymnasium)	Number of cases
Baden-Württemberg	48.9	27.8	23.3	240,507
Bavaria	53.2	20.8	26.0	205,666
Berlin	30.7	34.7	34.5	73,137
Brandenburg	37.5	27.1	35.3	4,296
Bremen	32.6	29.3	38.1	15,987
Hamburg	42.2	28.6	29.1	58,975
Hesse	35.7	34.0	30.4	128,353
Lower Saxony	17.0	45.4	37.7	6,815
Mecklenburg	44.6	29.9	25.6	87,406
North Rhine-Westphalia	48.0	26.4	25.6	330,665
Rhineland-Pfalz	42.7	27.4	29.9	64,225
Saarland	46.4	18.0	35.6	49,082
Saxony	15.0	22.4	62.6	5,569
Saxony-Anhalt	25.2	24.9	49.9	7,530
Schleswig-Holstein	42.3	28.5	29.2	55,112
Thuringia	0.0	37.8	62.3	2,656
Germany	45.1	27.2	27.7	1,335,980
West	46.5	26.7	26.8	1,235,978
East	28.0	33.8	38.2	100,002

Source: Scientific use files of the German Microcensus 1998–2005, own estimations
Notes: The educational attainment category "missing", which contains the sub-categories "no information" and "no degree", is omitted. "Foreigners" include EU-Foreigners and Non-EU-Foreigners. Weighted to national level using sampling weights provided with the German Statistical Office. Pearson chi-squared test: p = 0.000

On the contrary, in East Germany, the number of graduates is projected to decline on all educational levels (see Fig. 6.13). Yet, this reduction will be most notably in institutions of middle level education. By 2030, the number of graduates at the lower and middle level of education is projected to amount to less than 30 % of the base level in 2000, while the number of graduates with higher education is projected to be cut in half. At the same time, the number of students without a formal school leaving certificate is projected to decline.

6.4   Trends in Health Status

Along with educational attainment, the physical and mental health status is defined as a dimension of an individual's human capital endowment (Becker 1964: 7–36; Grossman 1972: xv). Despite continuing growth in military technology, it is expected that the military profession will continue to place high physical demands on the soldier (Rohde et al. 2007: 138). A multitude of military tasks requires a high degree of fitness and fine motor skills, i.e. physical strength, endurance, coordination, and agility. In particular, the lifting and carrying of heavy equipment

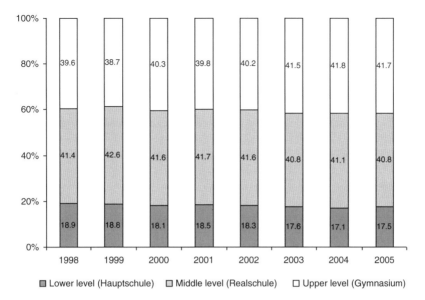

Fig. 6.9 Educational attainment, 1998–2005, Females 19–22 years, German (Source: Scientific use files of the German Microcensus 1998–2005. Notes: The educational attainment category "missing", which contains the sub-categories "no information" and "no degree", is omitted. Weighted to national level using sampling weights provided with the German Statistical Office. Pearson chi-squared test: p=0.000)

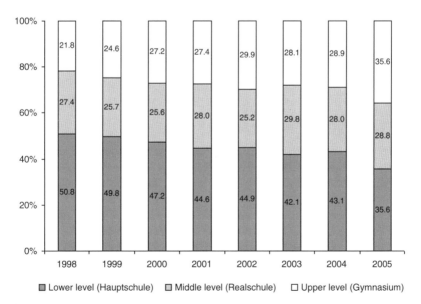

Fig. 6.10 Educational attainment, 1998–2005, Females, 19–22 years, Foreign (Source: Scientific use files of the German Microcensus 1998–2005, own estimations. Notes: The educational attainment category "missing", which contains the sub-categories "no information" and "no degree", is omitted. Weighted to national level using sampling weights provided with the German Statistical Office. Pearson chi-squared test: p=0.000)

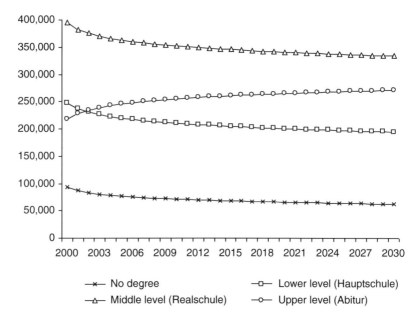

Fig. 6.11 Graduates by school type, Germany, 2000–2030, Log extrapolation (Source: KMK (2007), own estimations)

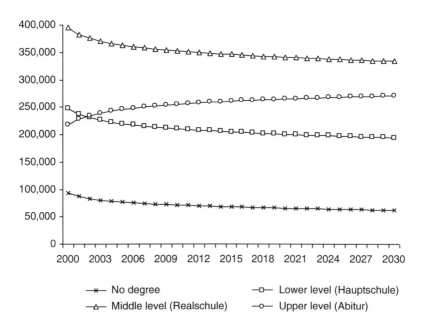

Fig. 6.12 Graduates by school type, West, 2000–2030, Log extrapolation (Source: KMK (2007), own estimations)

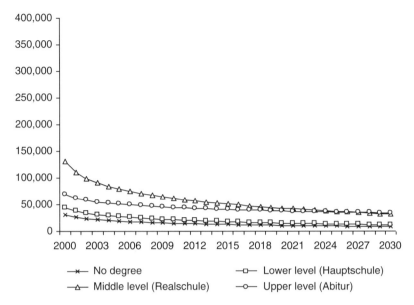

Fig. 6.13 Graduates by school type, East, 2000–2030, Log extrapolation (Source: KMK (2007), own estimations)

causes significant physical strain. During deployment, the physical requirements may be further exacerbated by extreme conditions in terms of sleep deprivation, supply shortfalls, climate, and rough terrain (Rohde et al. 2007: 138). Yet, evidence suggests an emerging discrepancy between manpower demand and supply. While the physical and mental requirements of soldiers have been on a continuous rise, recent trends indicate substantial health risks and deficits among youth.

The vast majority of youth in Germany, namely 85 % of the 10–16 year-olds (Ravens-Sieberer et al. 2003: 28) and 90 % of the 12–25–year-olds (Langness et al. 2006: 86) self-report a positive health status. Still, several critical problem areas related to youth health loom large. These include the early fixation of adverse health behaviors, such as substance misuse of alcohol or tobacco, sedentary life-styles with a lack of physical exercise, hypercaloric diet-induced obesity (Wabitsch et al. 2002: 251; Leyk et al. 2007: 143), deficits in fine motor skills (Opper et al. 2007: 886), and the increasing number of youth with allergic, chronic or mental conditions (Ravens-Sieberer et al. 2003: 31–48). Nearly 40 % of the 12–25-year-old youth population suffer from at least one allergy (41 % of females and 35 % of males). Males are more often affected by hay fever or asthma than females (Ravens-Sieberer et al. 2003: 33ff.). Evidence suggests that asthma affects the youth population more than any other age group. According to Boehmer et al. (2003: 55), about 15 % of the U.S. population aged 18–24 years are affected.

About 12 % of the 12–25 year-olds in Germany report a chronic disease that limits their activities of daily living. 50 % of the 12–25-year-olds indicate

Table 6.8 BMI classification
of body composition

BMI	Weight status
Below 18.5	Underweight
18.5–24.9	Normal weight
25.0–29.9	Overweight
30.0 and above	Obese

psychosomatic disorders. The prevalence of mental problems among the 12–25-year-old population ranged on a constant level of about 18 % in the past years, thereof 6 % self-report marginal mental conditions, while 12 % rate themselves as mentally troubled. In this connection, poor concentration, impulsivity, adaptive difficulties, hyperactivity, and conflicts with peers are reported most frequently (Ravens-Sieberer et al. 2003: 31–48). Boehmer et al. (2003: 58) suggest that, at least in the United States, the prevalence of mental disorders increased over the last decade.

6.4.1 Body Composition and Physical Capabilities

During the past 20 years, the prevalence of overweight has notably increased among children and adolescents, accounting for a wider spread of concomitant diseases of the respiratory, musculoskeletal, and metabolic system (e.g. high blood pressure, Diabetes Type II) among youth (Wabitsch 2004: 251ff.). The main predictors of weight gain and obesity include lifestyle behaviors related to diet and physical activity, along with socio-demographic characteristics, such as age, sex, family background, regional origin, or nationality.

The body mass index (BMI) constitutes the main prevalence measure of overweight and obesity in the population. It is calculated as weight in kilogram, divided by height in meters squared. The BMI classification distinguishes between four major categories, which are displayed in Table 6.8.

Research on the prevalence of overweight and obesity among children and adolescents, i.e. potential future military recruits, is limited due to the existing diversity in research methods, BMI threshold values, sample periods, reference groups, and the geographical focus (Wabitsch et al. 2002: 99f.; Apfelbacher et al. 2008: 128). However, two major trends emerge: Firstly, there has been a sharp increase in the prevalence of overweight and obesity among children and adolescents over time, in particular after the year 1985 (Kromeyer-Hauschild et al. 1998: 1145). Secondly, there are robust regional differences in the prevalence of overweight and obesity among children and adolescents, mainly arising from population heterogeneity. For example, there are marked regional differences in the composition of the general population as to nationality and socio-economic status (Wabitsch et al. 2002: 100). In addition, previous research demonstrated sex-specific variation in the prevalence and magnitude of overweight and obesity among children and youth, with young females mostly displaying a less favorable body composition (Kromeyer-Hauschild et al. 1998: 1145;

Table 6.9 Body composition, 1999–2005, Males 19–22 years, German

Body mass index					
	Percentage of sample				
Year	Underweight	Normal weight	Overweight	Obese	Number of cases
1999	2.7	76.5	17.4	3.4	1,177,116
2003	4.5	75.8	16.2	3.6	1,094,502
2005	4.3	74.0	18.4	3.3	1,302,324
Total	3.8	75.4	17.4	3.4	3,573,942

Source: Scientific use files of the German Microcensus 1999, 2003 and 2005
Notes: Weighted to national level using sampling weights provided with the German Statistical Office. Pearson chi-squared test: $p = 0.003$

Table 6.10 Body composition, 1999–2005, Males 19–22 years, foreign

Body mass index					
	Percentage of sample				
Year	Underweight	Normal weight	Overweight	Obese	Number of cases
1999	1.7	75.9	19.8	2.6	117,916
2003	4.4	63.6	27.9	4.1	120,109
2005	2.6	71.1	22.8	3.5	118,498
Total	2.9	70.2	23.6	3.4	356,523

Source: Scientific use files of the German Microcensus 1999, 2003 and 2005
Notes: Weighted to national level using sampling weights provided with the German Statistical Office. Pearson chi-squared test: $p = 0.203$

Wabitsch et al. 2002: 100). This is consistent with standard works of obesity research suggesting that young females are at a higher risk of overweight than young males (e.g. Daniels 2006: 73).

Despite the obvious flaws of the BMI as a measure for body status and physical performance, many Western militaries, including the Bundeswehr, use the BMI to determine the eligibility for service (Boehmer et al. 2003: 50). This significance of the BMI for military recruitment calls for a closer examination of respective trends among youth in Germany that belong to the primary target age group of the Bundeswehr.

Table 6.9 reports the trend in prevalence rates of overweight and obesity among males aged 19–22 years with German citizenship. Over the period between 1998 and 2005, the share of males of normal weight declined from 76.5 to 74.0 %. This decline was attended by both an increase in the proportion of underweight males (+1.6 %) and overweight males (+1.0 %) with German citizenship. In the meantime, the share of males classified as obese remained constant. Given current recruitment regulations, these males would be disqualified for service in the Bundeswehr.

Turning to the body composition of males with foreign citizenship, they tend to have a less favorable BMI than their German peers. Irrespective of some volatility in the trends over time, the average share of normal weight males with foreign citizenship amounted to 70 %, while the share of those with overweight ranged at nearly 24 % on average (see Table 6.10).

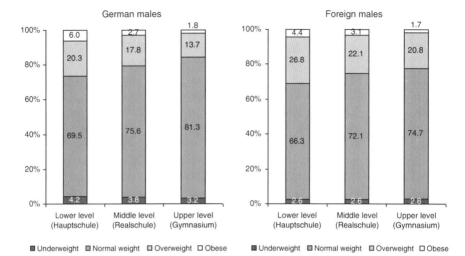

Fig. 6.14 Educational attainment and body composition, 1999–2005, Males (Source: Scientific use files of the German Microcensus 1999, 2003 and 2005. Notes: Weighted to national level using sampling weights provided with the German Statistical Office. Educational attainment categories "No information" and "No degree at time of interview" were omitted due to small number of cases and for consistency. Sample of German males: Pearson chi-squared test: p=0.000. Number of cases: Number of cases: 985,576 (Lower level); 1,384,899 (*Middle level*); 1,094,208 (*Upper level*). Sample of foreign males: Pearson chi-squared test: p=0.353. Number of cases: 152,259 (*Lower level*); 89,522 (*Middle level*); 78,102 (*Upper level*))

In view of the well-documented association between socioeconomic background and health status (e.g. Ravens-Sieberer et al. 2003; Langness et al. 2006; Murasko 2009), Fig. 6.14 reports the body composition of young males with German and foreign citizenship, separated by educational attainment. The latter is considered as a proxy of socio-economic status. In both subpopulations, there appear to be inverse educational gradients in overweight and obesity: Among the more highly educated, the share of overweight and obese males is smaller compared to the lower educated group.

However, the proportions of German and foreign males in each class of body composition range on different levels. The share of normal weight males with lower level education ranges at 70 % among those with German citizenship (66.3 % among foreign-born) and the respective share for males with middle and upper level education amount to 75.6–81.3 % (72.1–74.7 % among foreign-born).

Turning to the body composition of females with German and foreign citizenship, the relatively high prevalence rates of underweight are striking. With an average level of 11.9 % between 1998 and 2005, the share of German females affected by underweight even outranges that of overweight specified with 10.7 % (see Table 6.11). Over time, the proportion of normal weight German females declined

Table 6.11 Body composition, 1999–2005, Females, German

Body mass index					
	Percentage of sample				
Year	Underweight	Normal weight	Overweight	Obese	Number of cases
1999	12.5	75.2	9.5	2.8	1,059,667
2003	11.6	73.6	11.7	3.2	1,066,164
2005	11.7	73.9	10.7	3.7	1,174,968
Total	11.9	74.2	10.7	3.2	3,300,798

Source: Scientific use files of the German Microcensus 1999, 2003 and 2005
Notes: Weighted to national level using sampling weights provided with the German Statistical Office. Pearson chi-squared test: $p = 0.078$

Table 6.12 Body composition, 1999–2005, Females, foreign

Body mass index					
	Percentage of sample				
Year	Underweight	Normal weight	Overweight	Obese	Number of cases
1999	10.1	74.1	12.5	3.3	141,566
2003	9.7	76.2	9.0	5.1	120,910
2005	12.9	71.5	12.7	2.9	125,296
Total	10.9	73.9	11.5	3.8	387,771

Source: Scientific use files of the German Microcensus 1999, 2003 and 2005
Notes: Weighted to national level using sampling weights provided with the German Statistical Office. Pearson chi-squared test: $p = 0.624$

from 75.2 to 73.9 %. At the same time, the proportion of overweight and obese females increased by about 1 % each.

The body composition of females with foreign citizenship, as reported in Table 6.12 is nearly identical, albeit slightly lower average prevalence rates of underweight (10.9 %) and slightly higher proportions of overweight and obese females (11.5–3.8 %). Overall, however, with 73.9 %, the share of normal weight females with foreign citizenship is comparable to that of their German peers.

An examination of the body composition of females with German and foreign citizenship, separated by educational attainment, reaffirms the negative educational gradients in overweight and obesity already observed among males: Females with more education are less affected by overweight or obesity (see Fig. 6.15). Contrary to the presented findings for males, however, the body composition of females with German citizenship is not always more favorable than that of non-German females. Rather, on the lower and middle educational level, females with foreign citizenship feature a markedly better body composition in terms of a lower prevalence of underweight and overweight along with a higher share of females with normal weight. Conversely, at the upper educational level, the share of normal weight females with German citizenship is higher (79.7 % vs. 74.8 %).

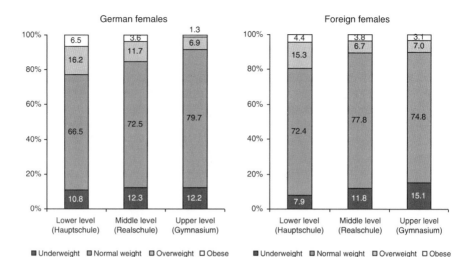

Fig. 6.15 Educational attainment and body composition, 1999–2005, Females (Source: Scientific use files of the German Microcensus 1999, 2003 and 2005. Notes: Weighted to national level using sampling weights provided with the German Statistical Office. The educational attainment category "missing", which contains the sub-categories "no information" and "no degree", is omitted. Sample of German females: Pearson chi-squared test: p=0.000. Number of cases: Number of cases: 566,199 (*Lower level*); 1,363,048 (*Middle level*); 1,299,936 (*Upper level*). Sample of foreign females: Pearson chi-squared test: p=0.013. Number of cases: 134,504 (*Lower level*); 99,139 (*Middle level*); 106,470 (*Upper level*))

6.4.2 Smoking Behavior

The potentially adverse effects of smoking and misuse of other substances on soldiers' performance and endurance, and therefore military readiness, have been discussed in the literature (cf. Klesges et al. 2001; IOM 2009) and call for a closer examination of related trends among youth in Germany. While evidence suggests that since the beginning of the 1970s, the average rate of substance misuse declined, the share of young alcohol and tobacco users in Germany still ranges well above the international average. Previous research indicated that about 39 % of the 12–25 year-old population consume alcohol at least once a week (Langness et al. 2006: 89), and about 8 % of the 15-year-old population reported to consume cannabis between 3 to 39 times per year. They are reckoned "leisure time consumers", while another 3 % consume cannabis more often and classify as "permanent consumers" (Richter and Settertobulte 2003: 122; Nickel et al. 2008: 77f.).

Estimates of the smoking prevalence among youth differ. Langness et al. (2006: 89) suggest that 26 % of the 12–25 year-old population smoke daily, while another 12 % smoke occasionally. With respect to the 17-year-old population, Lampert and Thamm (2007: 602) report an aggregate smoking prevalence of about 42 % for both males and females. However, in the German Microcensus, this order of magnitude only finds at older ages. There, only 17.2 % of the 15–18-year-olds

Table 6.13 Smoking behavior, 1999–2005, Males, German

Smoking behavior				
	Percentage of sample			
Year	Smoker	Ex-smoker	Non-smoker	Number of cases
1999	43.6	3.8	52.6	1,343,920
2003	46.5	4.6	49.0	1,289,275
2005	44.5	5.5	50.0	1,436,720
Total	44.8	4.6	50.6	4,069,915

Source: Scientific use files of the German Microcensus 1999, 2003 and 2005
Notes: Weighted to national level using sampling weights provided with the German Statistical Office. Pearson chi-squared test: p=0.000

Table 6.14 Smoking behavior, 1999–2005, Males, foreign

Smoking behavior				
	Percentage of sample			
Year	Smoker	Ex-smoker	Non-smoker	Number of cases
1999	42.33	6.04	51.63	140,451
2003	43.96	6.52	49.52	148,327
2005	42.38	5.23	52.39	131,727
Total	42.92	5.96	51.12	420,505

Source: Scientific use files of the German Microcensus 1999, 2003 and 2005
Notes: Weighted to national level using sampling weights provided with the German Statistical Office. Pearson chi-squared test: p=0.800

categorize themselves as smokers (16.1 % females and 18.2 % males) and 12.6 % thereof as regular smokers (12.0 % females and 13.2 % males) (Federal Statistical Office 2008a). The share of regular smokers is highest among those aged 20–24 years. In this age group, 38 % of males and 30 % of females consume tobacco regularly. Starting at age 40, the share of smokers declines steadily (Federal Statistical Office 2006b: 61f.).[2]

Turning to the prevalence of cigarette smoking among males in the Bundeswehr's primary target age group of 19–22 years, Table 6.13 suggests that the share of non-smokers among males with German citizenship somewhat declined between 1999 and 2005 (52.6 % vs. 50.0 %). In regard to males with foreign citizenship (see Table 6.14), it becomes evident that their smoking behavior is a bit more favorable than that of males with German citizenship. Their share of smokers is smaller (42.9 % vs. 44.8 %), and, regardless of ex-smokers, their share of non-smokers is slightly higher (51.1 % vs. 50.6 %).

---

[2] In the past few decades, the average age at smoking initiation continuously declined. In 2005, females and males of 65–68 years of age self-reported an average age at smoking initiation of 21.6 years and 17.8 years respectively. In the same year, surveyed youth in the age group of 20–24 years specified 15.2 years as their average age at smoking initiation (Federal Statistical Office 2006b: 61 f.).

Table 6.15 Smoking behavior, 1999–2005, Females, German

Smoking behavior				
	Percentage of sample			
Year	Smoker	Ex-smoker	Non-smoker	Number of cases
1999	36.2	5.2	58.6	1,059,667
2003	38.8	7.6	53.6	1,066,164
2005	37.6	7.2	55.2	1,174,968
Total	37.6	6.7	55.8	3,300,798

Source: Scientific use files of the German Microcensus 1999, 2003 and 2005
Notes: Weighted to national level using sampling weights provided with the German Statistical Office. Pearson chi-squared test: p = 0.000

Table 6.16 Smoking behavior, 1999–2005, Females, foreign

Smoking behavior				
	Percentage of sample			Number
Year	Smoker	Ex-smoker	Non-smoker	of cases
1999	28.47	5.61	65.91	141,566
2003	30.72	6.83	62.45	120,910
2005	25.14	4.64	70.22	125,296
Total	28.1	5.68	66.23	387,771

Source: Scientific use files of the German Microcensus 1999, 2003 and 2005
Notes: Weighted to national level using sampling weights provided with the German Statistical Office. Pearson chi-squared test: p = 0.113

The relatively healthier lifestyle of foreign males in terms of their smoking behavior reappears when looking at their female counterparts (see Tables 6.15 and 6.16). Over the period 1998–2005, the average share of non-smokers among German females was markedly lower than that of foreign females (55.8 % vs. 66.2 %). Simultaneously, trends over time differ. While among German females, the share of non-smokers declined, it *increased* among females with foreign citizenship.

As observed for the prevalence of overweight, there appears to be an inverse relationship between educational attainment and smoking (see Fig. 6.16). In particular, males with German citizenship feature a strong gradient between education and smoking behavior. Accordingly, while more than 60 % of German males with lower educational attainment self-report tobacco use, the respective share among more highly educated males is just about 27 %. The same gradient is observable for males of foreign citizenship, yet it is not as sharp, meaning that the differences in smoking behavior by level of educational attainment are less pronounced compared to their German counterparts.

In comparison with males, females feature lower smoking prevalence rates at all educational levels (see Fig. 6.17). When females of German citizenship are concerned, the share of smokers among those with lower educational attainment ranged

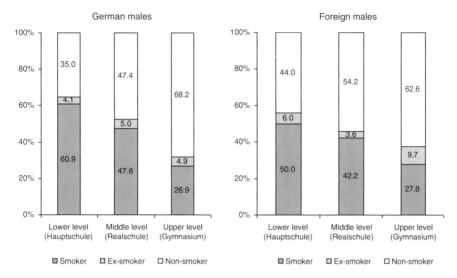

Fig. 6.16 Educational attainment and smoking behavior, 1999–2005, Males (Source: Scientific use files of the German Microcensus 1999, 2003 and 2005, own estimations. Notes: Weighted to national level using sampling weights provided with the German Statistical Office. The educational attainment category "missing", which contains the sub-categories "no information" and "no degree", is omitted. Sample of German males: Pearson chi-squared test: p=0.000. Number of cases: 1,126,317 (*Lower level*); 1,562,956 (*Middle level*); 1,248,489 (*Upper level*). Sample of foreign males: Pearson chi-squared test: p=0.472. Number of cases: 185,715 (*Lower level*); 10.2147 (*Middle level*); 87,450 (*Upper level*))

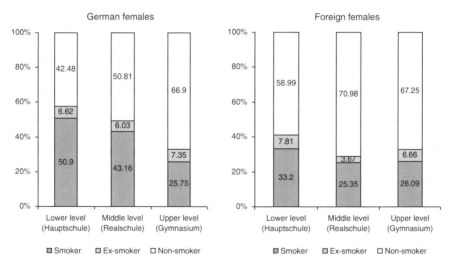

Fig. 6.17 Educational attainment and smoking behavior, 1999–2005, Females (Source: Scientific use files of the German Microcensus 1999, 2003 and 2005, own estimations. Notes: Weighted to national level using sampling weights provided with the German Statistical Office. Educational attainment categories "No information" and "No degree at time of interview" were omitted due to small number of cases and for consistency. Sample of German females: Pearson chi-squared test: p=0.000. Number of cases: 566,199 (*Lower level*); 1,363,049 (*Middle level*); 1,299,936 (*Upper level*). Sample of foreign females: Pearson chi-squared test: p=0.051. Number of cases: 134,504 (*Lower level*); 99,139 (*Middle level*); 106,470 (*Upper level*))

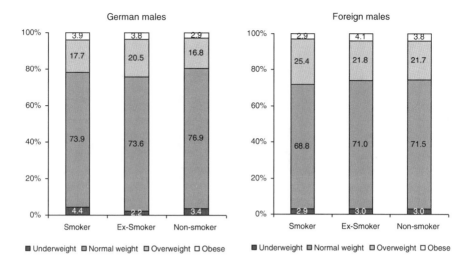

Fig. 6.18 Combined health risks (%), 1999–2005, Males (Source: Scientific use files of the German Microcensus 1999, 2003 and 2005. Notes: Weighted to national level using sampling weights provided with the German Statistical Office. Sample of German males: Pearson chi-squared test: p=0.000. Number of cases: 1,638,934 (*Smoker*); 168,169 (*Ex-smoker*); 1,737,844 (*Non-smoker*). Sample of foreign males: Pearson chi-squared test: p=0.578. Number of cases: 152,084 (*Smoker*); 24,219 (*Ex-smoker*); 175,781 (*Non-smoker*))

around 51 %. Of the more highly educated females, just about 26 % self-reported to be smokers. For females with foreign citizenship, the negative gradient between educational attainment and smoking behavior is not as pronounced. Foreign females with lower educational attainment level smoke at the highest rate (33.2 %). The share of female smokers with middle and upper level education is almost identical and even slightly higher for the latter group (25.3 % vs. 26.1 %).

Turning to the overlap in health risks among males, some interesting relations emerge (see Fig. 6.18). In regard to males with German citizenship, it becomes evident that (1) male smokers feature the highest prevalence of underweight; (2) male ex-smokers are more often overweight than smokers and non-smokers; and (3) male non-smokers feature the highest share of normal weight persons. With respect to males of foreign citizenship, the link between smoking behavior and body composition is not as clear-cut. Among non-German smokers, the share of overweight males is by far the highest, and the body composition of ex-smokers and non-smokers is largely comparable.

As to the overlap in smoking behavior and body composition among females of German and foreign citizenship, it appears that (1) female smokers generally have the highest prevalence of underweight, (2) just as for males, the proportion of over-weight females is highest for ex-smokers; and (3) female non-smokers have the highest proportions of normal weight females (see Fig. 6.19).

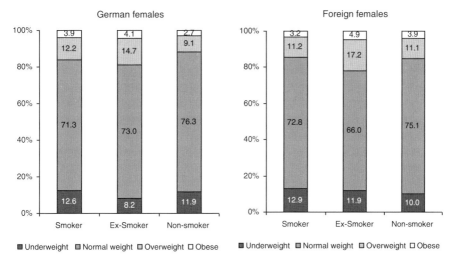

Fig. 6.19 Combined health risks (%), 1999–2005, Females (Source: Scientific use files of the German Microcensus 1999, 2003 and 2005, own estimations. Notes: weighted to national level using sampling weights provided with the German Statistical Office. Sample of German females: Pearson chi-squared test: p=0.000. Number of cases: 1,239,566 (*Smoker*); 220,742 (*Ex-smoker*); 1,840,491 (*Non-smoker*). Sample of foreign females: Pearson chi-squared test: p=0.822. Number of cases: 108,951 (*Smoker*); 22,017 (*Ex-smoker*); 256,803 (*Non-smoker*))

6.5 Trends in Employment Opportunities

The cohort size and the population age structure have been found to act as "a persistent and compelling feature of [an individual's] lifetime environment" (Ryder 1965: 844) and influence the occupational opportunity structure of youth. For example, Easterlin (1978: 401f.) argues that shifts in the size of cohorts[3] and population age structure, due to changes in fertility, affect the relative economic position and the social upward mobility of young males in various ways. Under the assumption of a constant rate of growth in the demand for younger and older workers in the labor market, a decline in fertility will lead to a relative scarcity of younger male workers, which will benefit their relative wages, unemployment risks, labor force participation rates, and their chances in the marriage market. With a smaller youth cohort size and the retirement of older workers, the educational and occupational opportunity structure of young males will improve considerably, solely on the basis of demographic change (Easterlin 1978: 401).

---

[3] According to demographic theory, a cohort is defined "as the aggregate of individuals (within some population definition) who experienced the same event within the same time interval" (Ryder 1965: 845). Accordingly, cohorts can be defined according to major events in the cohort lifecycle like simultaneously passing through an educational establishment organized by academic year.

Table 6.17 Changes in employment by sector, 2006–2015, Germany

	Levels (000s)				Δ 2006–2010		Δ 2010–2015	
	1996	2006	2010	2015	000s	%	000s	%
Δ in employment, all industries	37,270	38,095	39,014	40,068	918	2.4	1,054	2.7
Δ in employment, primary sector and utilities	1,541	1,197	1,063	953	−134	−11.2	−110	−10.4
Δ in employment, manufacturing	8,212	7,400	7,637	7,285	237	3.2	−352	−4.6
Δ in employment, construction	3,126	2,139	2,087	2,024	−52	−2.4	−64	−3.0
Δ in employment, distribution and transport	9,326	9,482	9,645	9,758	162	1.7	113	1.2
Δ in employment, business and other services	6,748	9,005	9,452	10,367	447	5.0	915	9.7
Δ in employment, non-marketed services	8,317	8,871	9,129	9,682	258	2.9	553	6.1

Source: Cedefop (2008: 67–73)

In view of the changes in the broader economic environment, the opportunity costs of military service will particularly increase for high-quality youth. Forecasts on labor market skill needs in Germany suggest that until 2015, sectoral changes in employment will lead to a major increase of personnel demand in the business and service sector and non-marketed services. Employment for less qualified personnel will decrease substantially, most notably in the primary sector, and, to a lesser extent, in the manufacturing and construction sector (see Table 6.17).

Table 6.18 represents the current occupational structure in Germany and illustrates how it will likely change between 2006 and 2015.[4] Noteworthy is the strong economic dependence on technicians, professionals, clerks, service workers, and craft-related workers. The projected change until 2015 indicates a considerable increase in employment in elementary occupations, as well as among service work-

---

[4] Cedefop (2008: 113) classifies the major occupational groups as follows: Armed forces; Legislators, senior officials and managers = legislators and senior officials, corporate managers, managers of small enterprises; Professionals = physical, mathematical and engineering science professionals, life science and health professionals, teaching professionals, other professionals; Technicians and associate professionals = physical and engineering science associate professionals; life science and health associate professionals, teaching associate professionals, other associate professionals; Clerks = office clerks, customer services clerks; Service workers and shop and market sales workers = personal and protective services workers, models, salespersons and demonstrators; Skilled agricultural and fishery workers; Craft and related trades workers = Extraction and building trades workers, metal, machinery and related trades workers, precision, handicraft, craft printing and related trades workers; Other craft and related trades workers; Plant and machine operators and assemblers = stationary plant and related operators, machine operators and assemblers, drivers and mobile plant operators; Elementary occupations = sales and services elementary occupations, agricultural, fishery and related laborers, laborers in mining, construction, manufacturing and transport.

Table 6.18 Status quo and projected change in major occupational groups

Occupation major group	Armed forces	Legislators, senior officials, managers	Professionals	Technicians, associate professionals	Clerks	Service workers, shop/market sales workers	Skilled agricultural/ fishery workers	Craft/ related trades workers	Plant/ machine operators, assemblers	Elementary occupations	All
Occupational employment, 2006 (000s)	193,000	2,162,000	5,025,000	7,963,000	4,687,000	5,124,000	671,000	5,639,000	2,602,000	4,031,000	38,095,000
Occupational structure, 2006 (row %)	0.5	5.7	13.2	20.9	12.3	13.4	1.8	14.8	6.8	10.6	100.0
Projected employment change by occupation, 2006–15 (000s)	8,000	–83,000	518,000	325,000	–576,000	856,000	–144,000	135,000	–36,000	967,000	1,972,000
Projected employment change by occupation, 2006–15 (%)	4.3	–3.8	10.3	4.1	–12.3	16.7	–21.4	2.4	–1.4	24.0	5.2

(continued)

Table 6.18 (continued)

Occupation major group	Armed forces	Legislators, senior officials, managers	Professionals	Technicians, associate professionals	Clerks	Service workers, shop/market sales workers	Skilled agricultural/ fishery workers	Craft/ related trades workers	Plant/ machine operators, assemblers	Elementary occupations	All
Replacement demand by occupation, 2006–15 (000s)	63,000	543,000	1,093,000	1,434,000	837,000	965,000	163,000	1,058,000	520,000	966,000	7,641,000
Total job openings by occupation, 2006–15 (000s)	71,000	461,000	1,611,000	1,759,000	260,000	1,821,000	20,000	1,193,000	485,000	1,933,000	9,613,000

Source: Cedefop (2008: 75–83)

ers and professionals. As a result, there will be an immoderate replacement demand for these occupations and a significant number of job openings to be filled with qualified personnel. It appears that even without demographic change, the competition for skilled labor would have become fiercer in the foreseeable future due to changes in Germany's economic structure and ensuing skill needs of the private sector. On the assumption that civilian employers will launch more attractive recruitment incentives to meet their quantitative and qualitative personnel requirements, it looms ahead that the relative competitiveness of the Bundeswehr as an employer will further decline as the share of the youth population contracts.

6.6   Trends in Youth Values and Attitudes

Parallel to the demographic shifts under way, there is indication of a sustained social change in Germany that affects individual values, preferences and life courses (cf. Ebenrett et al. 2001). In this context, value orientations are understood as stable and generally binding attitudes acquired during the process of socialization and that serve as a benchmark of behaviors and perceptions (Klages and Gensicke 2005: 279). Value orientations strongly influence the development of aptitudes, attitudes and aspirations.

Value change, in terms of shifted preferences, and the universal trend towards individualization have ranked high in the sociological discourse over the past decades. Based on the works of Inglehart (1971, 1977) about the social devaluation of existential needs and physical security in times of material prosperity, the prevalent view has been that "post-material" values like esteem, affiliation, participation and self-fulfillment have come to the fore in people's aspirations. However, more recent research on youth's value system emphasize that conventional, materialist values and the pursuit for self-actualization do not exclude each other. Rather, instead of substituting materialist and post-materialist values, today's youth appear to synthesize both (Gensicke 2002: 155ff.).

Research grouped the value orientations of German youth into several distinct categories (cf. Gensicke 2006). In particular, the groupings "private harmony" (i.e. friends, family), and "individuality" (i.e. independence, room for creativity) are highly valued. In contrast, youth approve the least those values associated with "tradition and conformity", such as holding on to established facts and procedures or doing what other people do (Gensicke 2006: 178–181). Another value category important to youth are "secondary virtues" including the respect for law and order, the pursuit of material security, diligence, a purpose in life and tolerance (Gensicke 2006: 190). This renaissance of "secondary virtues" originated in another value change during the 1990s that entailed a renunciation from the dominant orientation towards self-fulfillment and power as the two main priorities in life (Hurrelmann et al. 2006: 39f.).

The impact of social change on the Bundeswehr is multi-faceted. It manifested itself in the constantly high rates of conscientious objection until the abolishment of conscription in 2011, and continues to impede the recruitment of career personnel. This distaste for military service may be based on various grounds, which are either: (1) purely instrumental based on the view of the military as an inadequate concept of life; (2) socially driven based on the dislike of family and friends; or (3) morally based due to the incompatibility of ethical values and bearing arms (cf. Kohr 1996).

The differentiated value orientation of youth is equally reflected in their life and career planning. According to a study by Schaffer (1992), about 46 % of young males favor a balanced concept of life, in which the professional career and the partnership (or family) are equally important. A life concept that is purely oriented towards either leisure time, self-fulfillment or the family has lower approval rates (15.0–16.4 %). Since starting a family of their own plays an important role in the life concept of youth, they align their educational and professional development with the longer-term compatibility of professional and private aspirations (Schaffer 1992: 16).

Against this background, it can be assumed that, in the foreseeable future, the attractiveness of the soldier profession will further decline due to certain imminent characteristics. These include the risks to life and limb, long absences from home, high demands on the geographic mobility of soldiers and their families, as well as a high degree of bureaucracy and rigidity (cf. Segal 1986). These occupational disadvantages are further exacerbated by three trends in contemporary Germany: (1) modified role models in modern partnerships and the mutual pursuit of professional self-realization; (2) the regional demographic disparity and the social, economic and infrastructural degradation in the immediate vicinity of many Bundeswehr sites, which renders the integration of military families and working spouses ever more difficult; and (3) the divide between the preferred applicant profile and the – for such a personality type – inadequate prospects of self-determination, participation and career development in the Bundeswehr.

According to Kohr (1996), social change has further implications for the Bundeswehr and the viability of military service. He argues that foreign military deployments will continue to cause societal agitation and that the institutional orientation of the Bundeswehr towards traditional morals based upon comradeship, group cohesion, and obedience are contrary to the civilian system of values and thereby alienate youth that, in view of their cognitive and physical endowments, would be best suited for contemporary soldierly requirements. To that effect, youth have expressed unease with what they perceive as "uniformization, de-individualization, and de-personalization" during military service (Biehl and Kümmel 2003: 186).

Against this background, it seems worthwhile to draw upon some general observations of youth's views about the Bundeswehr as a place to work and their intention to serve as basic a military recruit or professional soldier. Previous research demonstrated that military propensity was predictive of actual military enlistment (e.g. Woodruff et al. 2006; Orvis et al. 1996; Bachman et al. 1998), which is why the following section explores military plans among several sub-populations of youth.

In a brief empirical analysis of perceptions about the military as a workplace, three survey items are considered: In terms of youth's interest in the Bundeswehr as a potential employer, possible answer categories include "interested", "less interested" and "not interested". When youth were asked about their plans for basic

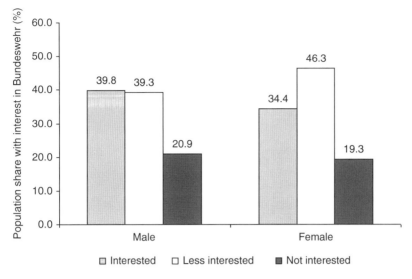

Fig. 6.20 Interest in Bundeswehr as potential employer, males and females (Source: AIK (2006), own estimations. Notes: Weighted to national level. Pearson chi-squared test: p=0.005. Sample size: 1,996)

military service or a military career, they could specify positive propensity, negative propensity or being "undecided". Similar to the concept of a "latent" military recruitment potential introduced by Bulmahn (2007: 39), it is assumed that individuals "less interested" in a military career or "undecided" are still susceptive to recruitment incentives of the military and are therefore, not a priori lost to other major institutions of higher education or civilian career development.

Figure 6.20 reports gender differences in youth's interest in the military as a potential employer. About 40 % of males and 35 % of females stated interest in the Bundeswehr as a place to work. The share of youth not interested in the Bundeswehr as a potential employer is well-balanced between males and females and ranges around 20 %. The remaining share of youth is less interested in the Bundeswehr but does not rule out a military career in the first place.

An examination of regional differences in military plans suggests that youth from East Germany are prone to military service at consistently higher rates than youth from West Germany, regardless of whether they are surveyed about the Bundeswehr as a workplace, basic military service or a longer-term military career. Accordingly, Fig. 6.21 reports that about 47 % of East German youth consider military service as a place to work compared to 34 % of their peers in West Germany. Conversely, the share of youth that preclude military service in the first place is higher in West Germany (21.8 %) than in East Germany (14.7 %).

Similarly, with about 49 %, the share of East German males interested in basic military service instead of the alternative civil service is about 10 % greater than in West Germany, whereas the proportion of males that is still undecided about their

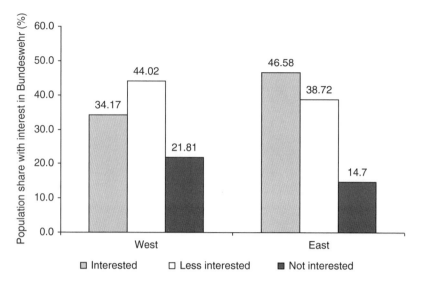

Fig. 6.21 Interest in Bundeswehr as potential employer, East vs. West (Source: AIK (2006), own estimations. Notes: Weighted to national level. Pearson chi-squared test: p=0.000. Sample size: 1,996)

preferred type of national service is at a similar level of about 17 % (see Fig. 6.22). In view of males' interest in the military as a profession (see Fig. 6.23), the differences between East and West are somewhat smaller (40.9 % vs. 34.9 %). The proportion of undecided males is markedly smaller in West Germany (10.6 %) than in East Germany (14.6 %). Therefore, there is a smaller population of targetable and potentially susceptible males in West Germany.

There are also strong distinctions by educational attainment to enlist as a basic conscript or professional soldier. Figures 6.24 and 6.25 reveal an inverse relationship in each case. Relating to basic military service, the share of interested males is larger at the lower secondary level, i.e. at an earlier stage of their educational development, compared to subsequent levels of education, such as upper secondary or tertiary education. The share of undecided males ranges between 13.2 % among males at the upper secondary level and 22.7 % among males at the post-secondary level (see Fig. 6.24). When a military career is concerned, the share of undecided males is largest at the lower secondary level, i.e. an earlier stage of the educational development. In equal measure, the share of males with plans for basic military service decreases as educational attainment increases (see Fig. 6.25). There is a clearly negative military propensity and therefore a very limited recruitment potential among university students or graduates.

This inverse relationship between educational attainment and military plans appears reasonable given that most students at the university have already decided upon a specific career. To a lesser extent, this also applies to the military plans of students at the upper secondary level. In equal measure, it is comprehensible that a

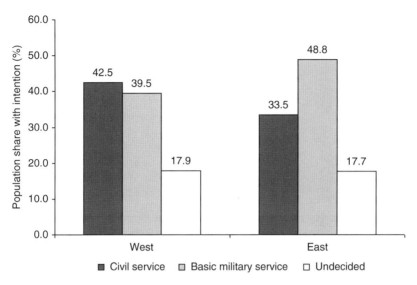

Fig. 6.22 Intention, basic military service, males, East vs. West (Source: AIK (2006), own estimations. Notes: Weighted to national level. Pearson chi-squared test: p = 0.037. Sample size: 845)

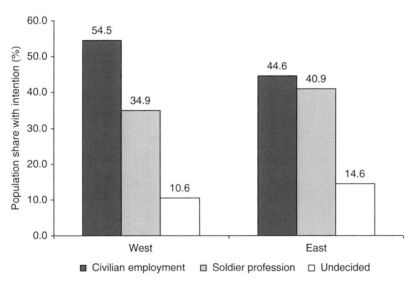

Fig. 6.23 Intention, soldier profession, males, East vs. West (Source: AIK (2006), own estimations. Notes: Weighted to national level. Pearson chi-squared test: p = 0.196. Sample size: 365)

relatively large share of males at the post-secondary level considers basic military service or a professional military career. Possibly, they intend to use their vocational skills as a door opener into the military.

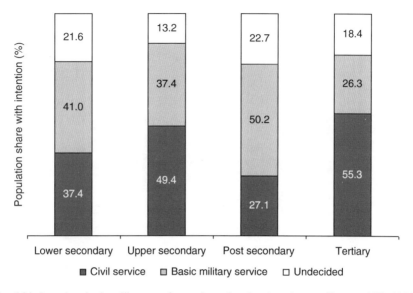

Fig. 6.24 Intention, basic military service, males, educational attainment (Source: AIK (2006), own estimations. Notes: Weighted to national level. Pearson chi-squared test: p=0.000. Sample size: 845)

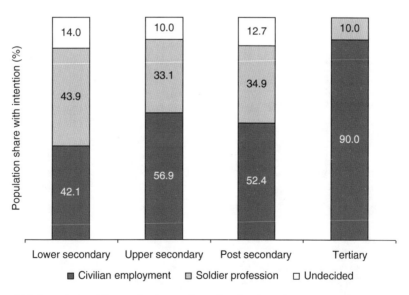

Fig. 6.25 Intention, soldier profession, males, educational attainment (Source: AIK (2006), own estimations. Notes: Weighted to national level. Pearson chi-squared test: p=0.080. Sample size: 365)

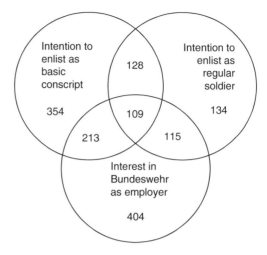

Fig. 6.26 Overlap of military plans (Source: AIK (2006), own estimations. Notes: Weighted to national level. Sample size: 1,014 males)

Separate analyses of youth's views about the Bundeswehr as a potential employer and the viability of basic military service or a military career are important to understand military propensity. At the same time, it may be relevant to interconnect all three of the variables to explore the overlap of interest and get more conclusive evidence of males' military plans.

Figure 6.26 displays the result of this exercise. It shows that, of the total sample of 1,014 males, 404 are interested in the Bundeswehr as a potential employer. Thereof, 213 intend to enlist as a basic military conscript and 115 plan to become a professional soldier. 128 males intend to enlist as a basic service recruit and as a professional soldier. In all likelihood, the most reliable estimate of the Bundeswehr's core recruitment potential consists in the 109 males that are interested in the Bundeswehr as a workplace and state their intention to enlist both as a basic military conscript and professional soldier. Therefore, just about 11 % of all males in the sample have a vested interest in military service as an occupational choice.

6.7 Trends in Parents' Educational Attainment

In line with standard economic theory, Becker (1981: 102) argues that income and fertility are inversely related, which is why "the effective price of children rises with income".[5] Given that couples prefer to invest their scare resources of time and

---

[5] As the main reason for the increased cost of children, Becker (1981: 102) cites the higher educational attainment of women, which entails a higher value of their time due to increased returns on their investments in education and other human capital enhancement. At large, the model suggests different fertility patterns based on parental income, human capital endowments, and cultural preferences.

money in the human capital of fewer children, couples with different levels of educational attainment and material resource endowments will bear children at different rates (Becker 1981: 93–112). As a consequence for the military, fluctuations in patterns of fertility across time and social strata will affect the enlistment propensities of subsequent youth cohorts and result in varying enlistment propensities among youth from differential family backgrounds (NRC 2003: 41).

Thus, along with personal attributes, family background factors play a major role in occupational choice and influence youth's propensity for military service (Warner et al. 2001: vii). In this regard, there are two major trends with adverse implications for military recruiting: The improvement in parent's education and the declining share of parents with a military background.

The described shift in parental choice from child quantity toward child quality is attended by a transmission of lifestyle preferences from parents to children in terms of welfare aspirations and vocational objectives. To that effect, Ryder (1965: 854) showed that children usually have the ambition to achieve a welfare level that is at least equal to the welfare level of their parents. With today's parents having a higher educational attainment than previous parental generations, they influence their offspring's aspirations for higher education and create an environment that is more intellectually stimulating, motivating and prosperous compared to previous cohorts (NRC 2003: 57). Given that the occupational preferences of youth are largely formed within the socio-economic environment provided by their parents, it comes with no surprise that propensity for military service tends to be lower among youth with better-educated parents (Warner et al. 2001: 26). In this context, the increasing share of children to mothers with at least some university education is considered of particular importance (NRC 2003: 59).

Table 6.19 illustrates the increased average educational attainment levels of the parental generation in Germany. On the assumption that parents of the current target population were between 50 and 55 years in 2007, the figures clearly indicate this group is more highly educated than persons over 65 years, regardless of which vocational degree is considered. In particular, the share of persons with a university degree or a degree from a university of the applied sciences increased. Given these improvements in parents' educational attainment, it is foreseeable that an increasing share of the current and future generation of children will be urged to pursue higher education.

6.8   Trends in Parents' Military Background

Family linkages to the military have been recognized as an influential factor in enlistment decision-making, in particular among high-quality youth (Faris 1981: 545). Previous research consistently demonstrated that sons (and daughters) of professional soldiers were severalfold more likely to choose a military career than their peers without military affiliation (e.g. Faris 1981: 550; Thomas 1984: 307). When children are raised by parents with a military background, they

Table 6.19 German population in 2007 by age and educational attainment

Age groups	Total	Vocational degree			University of applied sciences	University	Doctorate	No information	Without vocational degree
		Apprenticeship	Technical/ professional school	Technical/professional school in former GDR					
in 1,000									
Total	**66,382**	**36,023**	**3,987**	**829**	**3,223**	**4,844**	**676**	**171**	**16,269**
in %									
20–25	100	39.9	2.0	–	1.0	0.8	–	0.1	56.1
25–30	100	54.7	5.1	–	5.0	8.5	0.2	0.2	26.1
30–35	100	55.1	6.5	–	6.7	12.0	1.2	0.3	18.0
35–40	100	58.2	7.0	0.9	6.6	9.9	1.4	0.3	15.4
40–45	100	59.5	7.7	1.4	6.2	8.7	1.4	0.3	14.6
45–50	100	59.3	7.5	1.6	6.0	8.5	1.2	0.2	15.4
50–55	100	58.7	6.7	1.8	6.0	9.3	1.3	0.2	15.6
55–60	100	58.1	6.5	1.8	5.3	8.6	1.2	0.3	17.9
60–65	100	57.0	6.4	2.1	5.0	7.8	1.4	0.2	19.8
65+	100	49.0	5.1	1.6	3.2	4.3	0.9	0.3	34.2
Total	**100**	**54.9**	**6.0**	**1.6**	**5.1**	**7.8**	**1.1**	**0.2**	**23.3**

Source: Federal Statistical Office (2008b: 131)

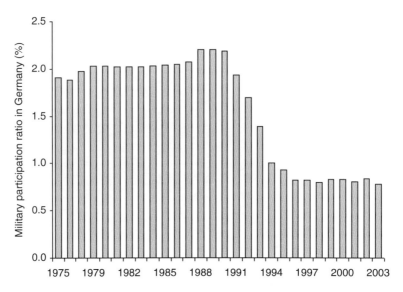

Fig. 6.27 Trends in military participation ratio in Germany (Source: IISS 1975–2003 (cited in Szvircsev Tresch 2005: 258))

receive exclusive information about the military lifestyle from their primary influencers, which will likely influence their occupational aspirations and expectations (NRC 2003: 60).

In light of the large-scale force reductions of the Bundeswehr since the end of the Cold War, the share of parents and other influencers who served in the Bundeswehr declined. The effect on military recruiting is two-fold: parents may be less supportive of a military career (Warner et al. 2001: 1), and a lower share of fathers acts as a soldierly role model for adolescent youth.

Figure 6.27 reports the evolution of the military participation ratio in Germany between 1975 and 2003, defined as the proportion of the national population registered by the military, including both active and reserve forces (Haltiner 1998: 12). Given the downward trend over time and further force reductions during the military reform initiated in 2011, an ever smaller share of the population will have military experience in the future. Family tradition as a potential recruitment driver will therefore be ever less significant.

6.9   Projection of Youth's Human Capital Until 2030

This section presents a modest attempt to assess the future human capital endowments of the most relevant recruitment potential of youth between 19 and 22 years, illustrating how the projected demographic changes may influence the availability of skilled and physically capable soldiers until the year 2030. The method draws upon official projections of the youth population along with information about trends in educational attainment and health status, as measured by smoking status and body mass index.

6.9.1   Data

6.9.1.1   12th Coordinated Population Projection

The 12th coordinated population projection covers a period until 2060 and provides several variants that fix the "limits of a corridor", within which Germany's population size and age structure will evolve given certain assumptions about future trends in fertility, mortality, and migration (Federal Statistical Office 2009). The 12th coordinated population projection, therefore, does not constitute a forecast in its original sense of predicting future demographic trends with absolute certainty. It rather provides basic future-oriented information about the German population and should be understood as a model of long-term demographic trends, whose predictive power declines as the projection period increases.

The demographic basis of the human capital projection consists in the lower limit "medium" population variant of the 12th coordinated population projection, as available from the Federal Statistical Office. It is based on three major assumptions: Firstly, the birth rate will remain nearly constant at 1.4 children per woman and the mean age at birth will rise by about 1.6 years. Secondly, life expectancy at birth will reach 89.2 years for females and 85.0 years for males in 2060. Thirdly, future net migration will gradually increase and level out at about 100,000 until 2060. In view of the increasing degree of uncertainty associated with future population change, the projection horizon is restricted to the year 2030.

6.9.1.2   Microcensus

The Microcensus is an annual cross-sectional survey of 1 % of German households with a response rate of about 97 % due to compulsory participation. It provides representative information about the demographic and socio-economic fabric of the German population, including insights into the level of educational attainment and type of employment (e.g. Shahla et al. 2005). There is an additional, regularly conducted survey program of a 0.45 % sample. In 1999, 2003 and 2005, it surveyed the health status of the respondents.

Information about the educational attainment and health status of the youth population is drawn from the Scientific Use Files of the years 1998–2005. They contain a 70 % random sample of the original 1 % household sample. In order to estimate human capital endowments of eligible youth at recruitment age, all analyses are restricted to males and females from 19 to 22 years of age with a school leaving certificate at the time of the interview. Given the lack of differentiation between the native and foreign-born resident population in the 12th coordinated population projection, non-native youth are likewise included in the Microcensus sample. All survey years are pooled in order to ensure sufficient representation of the narrow age range under review.

6.9.1.3   School Graduate Projection

The projection of school students and graduates is published regularly by the Standing Conference of the Ministers of Education and Cultural Affairs of the Federal States. Projected figures are available for East and West Germany, and the 16 German states separately (KMK 2007). They take into account changes in state-specific school legislation, such as the reduction of school years required for a university entrance certificate and the resultant double cohorts of school graduates entitled to enter higher education. Similarly, the projection results provide the empirical basis for educational policy, such as defining the future demand of teachers or the required number of places in vocational training and university degree programs.

The available model calculation of school students and graduates up to the year 2020 is a status-quo forecast. Current transition rates from school to further education are assumed to remain constant into the future. The projection dynamic results from the demographic changes throughout the projection period.

6.10   Method

The main objective of the model calculation is to provide estimates of future trends in educational attainment and health status among 19–22 year-old males and females, i.e. the sub-population group most relevant for military recruitment. The underlying approach draws upon an extrapolative method and a modification of the cohort-component approach. The extrapolative approach is applied on the assumption that the future will be a continuation of the past and that regularity determines the development of any patterns and trends (Booth 2006: 550). In this, the extrapolative approach is largely atheoretical and makes no use of exogenous variables or current knowledge about actual and prospective developments in areas relevant to the projection (Booth 2006: 550f.). The core element of the modified cohort-component approach is a division of the base-year population into different education groups, separately accounting for their health status and essentially treating them as separate sub-populations (cf. George et al. 2004: 571; Smith et al. 2001: 43ff.). Projections by educational attainment and health status are estimated for males and females in East Germany, West Germany and all Germany.

The baseline year 2010 provides the empirical starting point of the projection. Using the population distribution by single year groups from the 12th coordinated population projection, males and females at ages 19–22 years are summarized into one age group each. Since the coordinated population projection is a residential population projection, it does not distinguish citizenship status. Under the assumption that the majority of school students have graduated by the age of 19–22 years, the official school graduate projection is used to calculate the projected relative shares of four educational attainment levels, namely lower education (graduated from a general education secondary school, i.e. "Hauptschule"), medium education (graduated from a secondary or junior high school, i.e. "Realschule"), higher education (graduated from an academic high school, i.e. "Gymnasium") or no school-leaving certificate. These projected shares of educational achievement are applied to the absolute youth population projection estimates of males and females together. Given that the official school graduate projection only covers the period from 2005 to 2020, it is assumed that the 6-year average from 2015 to 2020 will prevail into the future. It is therefore applied as a constant rate to the projected youth population estimates for the period from 2020 to 2030. However, this scenario is somewhat pessimistic since it keeps the proportion of each education group constant, not allowing for further educational expansion or other contextual change. Since the official school graduate projection is not differentiated by sex, the average female-to-male ratio of the Microcensus surveys from 1998 to 2005 is applied as a constant rate to the projected youth population by educational attainment. An intermediate result of this status-quo projection is the projected number of male and female graduates between 19 and 22 years in four educational attainment categories in East Germany, West Germany and all Germany from 2010 to 2030.

In a second step, two health risks – smoking and weight status – are applied separately to the calculated projection estimates by age, sex and educational attainment. On the basis of data from the Microcensus surveys in 1999, 2003 and 2005, the prevalence rates of smoking and non-smoking, as well as normal weight and overweight were applied to the projected youth population. In the projection that includes the smoking status, it was assumed that the average prevalence rate of the years 1999, 2003 and 2005 would stay constant in the future. The same assumption was made for the projection of youth's weight status. In view of the main concern about the future recruitment potential and in particular, the number of youths capable to serve in the military, only the projection estimates for fit candidates are displayed, namely non-smokers and those with normal weight.

6.11 Results

This section presents the projection estimates for educational attainment among 19–22 year-old youth in Germany, East Germany and West Germany from 2010 to 2030. The projections are made for four levels of education. The education levels refer to the highest level of schooling attained by the age of 19–22 years. The four projected levels are: no degree, low education, medium education, and high education.

Figures 6.28, 6.29, 6.30, 6.31, 6.32 and 6.33 report the projected absolute change in males and females of 19–22 years in Germany, East Germany and West

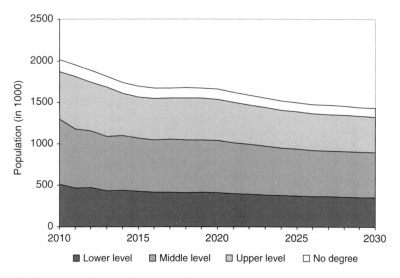

Fig. 6.28 Germany, males by educational attainment

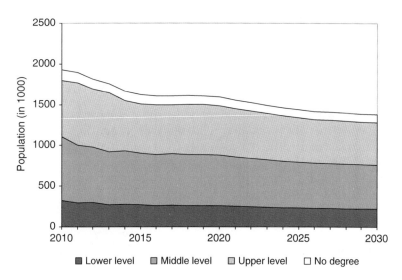

Fig. 6.29 Germany, females by educational attainment

Germany, separated by educational attainment. The height of all colored areas in each graph shows the size of the youth population over time. The different colors within the graphs indicate youth of different education levels. The top portion of each graph denotes youth with no school leaving certificate; the light gray portion below shows youth with high education, i.e. a university entrance diploma; the

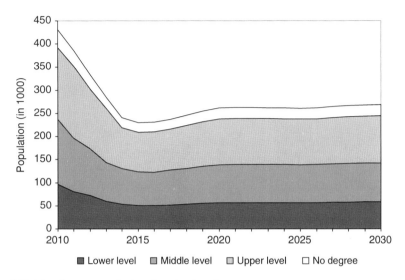

Fig. 6.30   East Germany, males by educational attainment

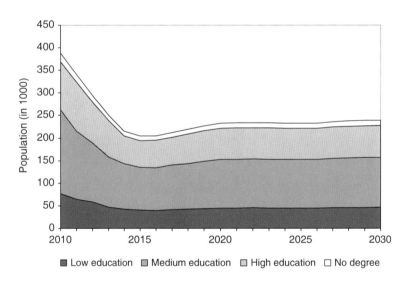

Fig. 6.31   East Germany, females by educational attainment

dark grey portion below shows youth with middle education, i.e. those qualified for vocational training; and the darkest portion at the bottom indicates youth with low education, i.e. those with a first general education. The population scale on the graphs varies and ranges from 450,000 for East Germany to 1.8 million for West Germany and 2.5 million for all Germany. Thus, the graphs are most useful for

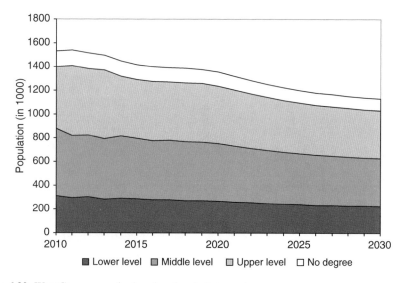

Fig. 6.32 West Germany, males by educational attainment

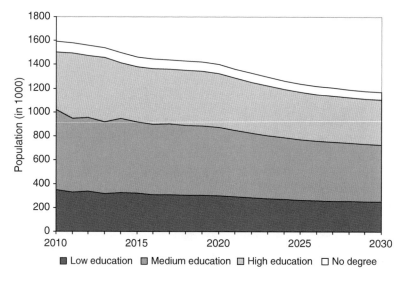

Fig. 6.33 West Germany, females by educational attainment

looking at the relative change of the youth population and the respective education groups from 2010 to 2030.

The projection estimates are mainly driven by demographic change and assumptions about future educational attainment rates. The youth populations in all regions are set to decline in absolute terms. Yet, the sweeping change projected for the youth population in East Germany is disproportionate. Particularly between 2010 and 2015, the population of young East German males (see Fig. 6.30) and females (see Fig. 6.31) will drop significantly. With this, all education groups will experience a major reduction in absolute numbers.

At the same time, there is a general trend towards higher levels of schooling. Thus, while demographic change reduces the number of school graduates (and school dropouts), a higher share of students pursues a degree that qualifies them for further education, either in vocational programs or at the university. This trend emerges among all sub-populations. For example, the relative share of young German males that complete medium education is projected to decline slightly by 2015, while the share of the high education group will increase accordingly (see Fig. 6.28). Thereafter, the relative distribution of education will remain stable (due to the underlying assumptions), while the absolute population size of young German males will decline. Among females, the relative shares of medium education and high education are higher than among males (see Fig. 6.28). However, the general trends are the same.

In East Germany, the population of youth between 19 and 22 years will almost halve by 2015. Compared to the shares for total Germany, fewer males in East Germany complete medium education (39 % vs. 33 %) and relatively more attain a higher level of schooling (28 % vs. 36 %). By 2020, the share of medium education is projected to somewhat decline, while the share of males with higher school education will increase accordingly (see Fig. 6.30). The starting population of East German females is noticeably smaller than that of males, supposedly due to out-migration already in late adolescence. About half of the remaining females between 19 and 22 years completed a medium level school education by 2010. Another 20 % attained a low school education. Both shares are projected to decline in favor of the high education group (see Fig. 6.31).

In West Germany, the population of males and females between 19 and 22 years is projected to decline less rapidly. In 2010, about 37 % of the males had completed a medium level school education and another 34 % a higher school education (see Fig. 6.32). The female population size is somewhat larger than that of males, most likely due to in-migration from East Germany. With a share of about 40 %, the majority of West German females completed a medium level school education. About 30 % attained a high level of schooling (see Fig. 6.33). Just like in other sub-populations, the trends among young males and females in West Germany follow the shift from a medium level school education to a higher level school education with a parallel decline the absolute number of graduates.

Figures 6.34, 6.35, 6.36, 6.37, 6.38 and 6.39 present estimates of the projected absolute change in the male and female youth population, separated by youth education level and weight status. The projection dynamic results from demographic change and assumptions about future developments in educational attainment and weight status. If normal weight status is added as a criterion for exclusion from the recruitment potential, its absolute size decreases, and the relative weight of education groups in the remaining population changes.

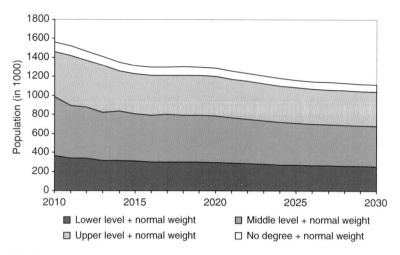

Fig. 6.34 Germany, males by educational attainment, normal weight

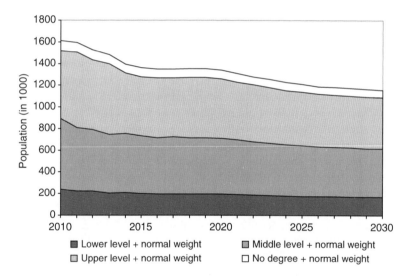

Fig. 6.35 Germany, females by educational attainment, normal weight

Hence, among young males in Germany, the inclusion of the normal weight variable reduces the absolute size of the recruitment potential and increases the relative share of young males with higher schooling (see Fig. 6.34). Concretely, if normal weight were installed as a necessary requirement for military enlistment, the relative share of young males with a low level education or without a school leaving certificate would shrink by a few percent. In parallel, the average educational attainment in the recruitment-relevant group would increase. That means, among males

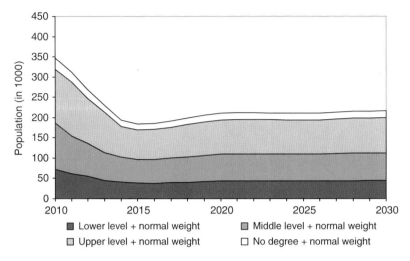

Fig. 6.36 East Germany, males by educational attainment, normal weight

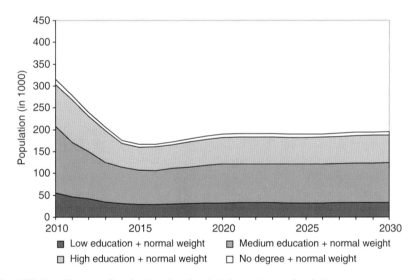

Fig. 6.37 East Germany, females by educational attainment, normal weight

of 19–22 years that are of normal weight and would therefore be qualified to enlist, a relatively higher share attained a university-entrance certificate compared to the total recruitment potential that disregards weight status. The same trends are observable among young males in East Germany (see Fig. 6.36) and West Germany (see Fig. 6.38). Over time, this trend is projected to become manifest. Thus, among

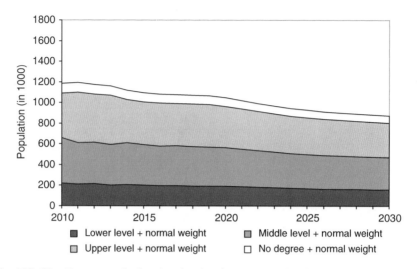

Fig. 6.38 West Germany, males by educational attainment, normal weight

males of normal weight status, the relative share of those with a high level school education will continue to increase until 2020 and then stagnate.

Similar to the effect among males, the inclusion of the normal weight criterion reduces the absolute size of the female recruitment potential in Germany and causes an upward shift in the distribution of education (see Fig. 6.34). Compared to the total population undifferentiated by weight status, the relative share of females with a low school education decreases, and the share of those with a high level school education increases. The same trends are observable among young females in East Germany (see Fig. 6.37) and West Germany (see Fig. 6.39). Until 2020, the share of lower educated females with normal weight will further decrease, while an increasing percentage in the recruitable population will be females with a university entrance certificate.

Figures 6.39, 6.40, 6.41, 6.42, 6.43, 6.44 and 6.45 present estimates of the projected absolute change in the male and female youth population, separated by youth education level and smoking status. If non-smoking is added as an alternative criterion for exclusion from the recruitment pool, its absolute size decreases, and the relative weight of education groups changes.

The inclusion of the non-smoking variable has a considerable effect on the size of the recruitment potential with a low school education. For example, in 2010, about 500,000 young males in Germany completed a low education school program by the age of 19–22 years. Taking into account the non-smoking criterion reduces the male recruitment pool to about 200,000. It also reduces the relative weight of the low education group from 25 % in the total recruitment population that disregards smoking status to 19 % (see Fig. 6.40). Hence, the absolute size of the non-smoking recruitment potential is much smaller, and a higher share of the non-smoking males

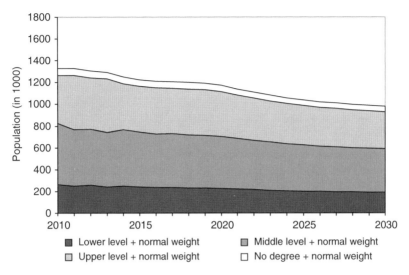

Fig. 6.39 West Germany, females by educational attainment, normal weight

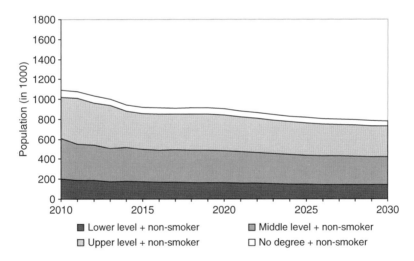

Fig. 6.40 Germany, males by educational attainment, non-smoker

has a school degree that opens many different paths for the future. The magnitude of the effect is similar for young males in East Germany (see Fig. 6.42) and West Germany (see Fig. 6.43). Over time, the share of non-smoking males with a low school education will continue to decrease, while the share of those that do not smoke and have a higher school education will grow accordingly.

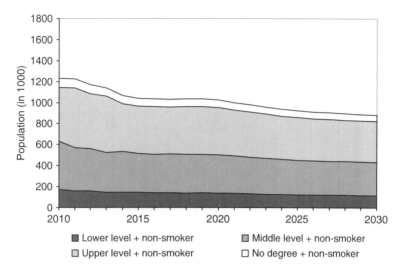

Fig. 6.41 Germany, females by educational attainment, non-smoker

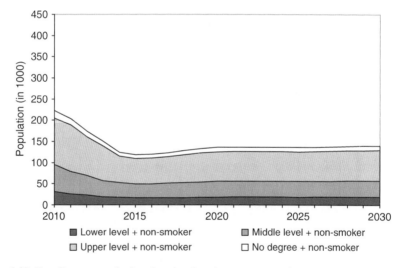

Fig. 6.42 East Germany, males by educational attainment, non-smoker

Among females, non-smoking as a criterion for exclusion from the recruitment potential equally has a reductive effect among those with a low school education but also among females with a medium school education. For example, about 320,000 females in Germany between 19 and 22 years had attained a low of schooling by 2010 (see Fig. 6.41). With the inclusion of the non-smoking variable, the respective recruitment pool declines to about 170,000, and the relative

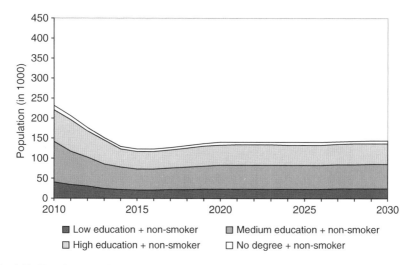

Fig. 6.43 East Germany, females by educational attainment, non-smoker

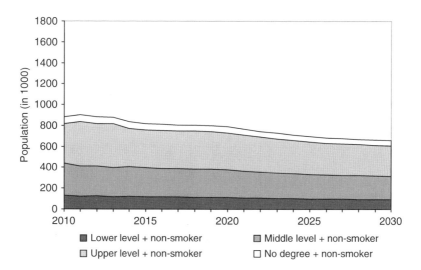

Fig. 6.44 West Germany, males by educational attainment, non-smoker

weight of females with a low level of education declines from 17 % in the total recruitment potential that disregards smoking status to 14 % among non-smoking females. In regard to German females with medium level schooling, about 780,000 had attained their degree by 2010. However, just under 460,000 remain in the recruitment pool if non-smoking were to be installed as a criterion of exclusion. The respective relative weight of medium education in the female recruitment

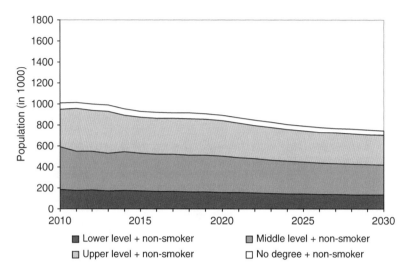

Fig. 6.45 West Germany, females by educational attainment, non-smoker

pool would decline from 41 % if non-smoking is disregarded to 37 % when non-smoking is an enlistment pre-requisite. Just as for the males, the inclusion of the non-smoking criterion would produce a better educated, albeit much smaller, recruitment pool. This trend is valid over time and region, i.e. East Germany (see Fig. 6.43) and West Germany (see Fig. 6.45).

6.12   Conclusion

Social changes, particularly in the level of education, will be a major obstacle to the recruiting success of the Bundeswehr in the future. Trends in the distribution of educational attainment among youth, as well as changes in the absolute weight of education groups, hint at a compounding recruitment challenge and should provide an indication of necessary changes in recruitment policy. The overall upward shift in educational attainment and the respective decrease of youth with a low or medium education may enforce the competition between the Bundeswehr and universities or providers of attractive vocational programs.

The picture becomes even gloomier when normal weight or non-smoking status is factored into the estimation of the recruitment potential. Considered as proxies for a healthy lifestyle and adequate physical performance, their inclusion substantially reduces the absolute recruitment pool of both males and females, mainly at the lower and medium level of schooling, and therefore increases the educational attainment of the remaining group. Hence, youth that are of normal weight or do not smoke, are on average better educated.

In the future, the relative share of well-educated youth will continue to increase, and so will the share of well-educated youth with a favorable health status. In the meantime, the absolute number and relative share of young males and females with a low or medium education will decrease. The reduction of the recruitment potential among these subpopulations will be even greater if the Bundeswehr installed normal weight or non-smoking as requirements for enlistment.

The projections highlight a recurring dilemma that the Bundeswehr will have to find solutions for in the near future: Those youth that are most capable of fulfilling the cognitive and physical requirements of the soldierly profession attain a level of schooling that predisposes them for a variety of career options, while the share of youth with a low or medium education will decline. This expansion in education may be favorable in terms of the recruitment of officers but will severely aggravate the recruitment situation among the lower ranks, i.e. the much needed troops on the ground. This is even more relevant when, besides educational attainment, good health is made a knock-out criterion for the recruitment of non-commissioned officers or enlisted personnel.

Chapter 7
Demographic Change as a Strategic Constraint: Issues and Options

7.1 Summary of Findings

Seeing the redefined international military role, as well as the neglect of both demographic research and military sociology in the past decades, Germany serves as a case study to outline the impact of demographic change on strategic calculations. The basic assumption of the thoroughly interdisciplinary research approach is that the Bundeswehr operates in an uncertain environment in terms of the quantity-quality dimension of future manpower demand and manpower supply. This raises the question about the determinants of adequate military manpower supply and how these will change due to demographic change. A set of potentially contributing factors includes young adult demography, educational attainment, health and physical attributes, youth's values and aspirations, family background and civilian employment opportunities. These factors, representing the supply side of military manpower, form the broad framework for the structure of the analysis. The international security environment and the ensuing requirements for Western military organizations and their soldiers form the demand side of military manpower.

The broader military environment defines military missions and the military organization. It comprises four main variables: (1) The process of globalization, as being reflected in an increased global connectedness in political, economic, military and cultural relations, leads to a heightened sensitivity to security-relevant developments worldwide. (2) The geostrategic environment yields a sustained high number of regional conflicts that simultaneously display low- and high-intensity elements of warfare. (3) Technical-engineering innovations, gathered under the term "revolution in military affairs", that are largely driven by social and political factors with profound implications for the character of warfare. (4) The socio-cultural environment, which is a vital source of the foreign policy orientation of industrialized states. It may either enable or constrain the government's capacity to carry out foreign policy and, if necessary, project military power. In Germany, there appears to exist a general anti-militarism in society, which may explain the sustained reluctance to have a greater share in the management of international security albeit

W. Apt, *Germany's New Security Demographics: Military Recruitment in the Era of Population Aging*, Demographic Research Monographs, DOI 10.1007/978-94-007-6964-9_7, © Springer Science+Business Media Dordrecht 2014

increasing external pressures, and a relative social marginalization of the military and the military profession. As a result of these changes in the external environment, the role set and organizational structure of the military in the twenty-first century are undergoing a major transformation. Western armed forces have been endorsed with an "extended role set", in which newly emphasized demands are complementary to the traditional roles aimed at the security and integrity of national territory. In response to the contemporary role set, there has been a trend towards military professionalization and a greater reliance on well-trained, immediately deployable professional soldiers. The Bundeswehr, for a long time a "pseudo-conscript force", also partly entertained the draft as a "personnel reservoir" to complement the mostly professional military organization. In 2011, the Bundeswehr initiated the transition to a fully professional military.

Empirical analyses of socio-demographic and other traits of soldiers in the Bundeswehr, based on the German Microcensus, yielded two broad conclusions: (1) There is a distinct selectivity of military service in Germany, regardless of whether the group of basic military service conscripts or professional soldiers is concerned. Hence, the established findings suggest that the Bundeswehr is more East German, more rural and more educated than the general population, while differences in health-related factors remain inconsistent. (2) While the Bundeswehr recruits a disproportionate share of military personnel from East Germany, it enlists East German males (and females) with comparatively high human capital endowments. This phenomenon of "positive selectivity" into the military is consistent with previous research and has also been observed among the Black male population in the United States.

Additional multivariate regression analyses showed that factors, which influence youth's initial interest in joining the military, equally control subsequent occupational decision-making regarding military service. Two general conclusions can be drawn in regard to the selectivity of military service and the efficacy of conscription as a recruitment tool for high-quality personnel: (1) The comparison of estimation results by region suggests that males from West Germany that choose *not* to enter the Bundeswehr as a regular or professional soldier are generally more likely to be higher educated, originate from an urban setting, have a better body composition, stem from a well-to-do family, and have fewer siblings than those entering the Bundeswehr. The same comparison for males from East Germany reveals that those who *enter* the Bundeswehr as regular or professional soldiers are more likely to be higher educated, possess a better body composition, and have more siblings than those not entering the Bundeswehr. Thus, it appears that the Bundeswehr "skims" the high-quality male population in Eastern Germany. (2) Conscription as a recruitment tool for the preferred high-quality youth has been shown to be largely ineffective. The findings consistently demonstrate that primarily males with less favorable human capital traits were drawn into the Bundeswehr via the channel of conscription. Those with higher human capital endowments chose to enter the Bundeswehr as regular or professional soldiers instead. In the future, the most significant obstacles to the recruitment of quality personnel consist in the decline of the youth population in East Germany, where military propensity and military participation had thus far been high, and the general increase in educational attainment.

The decline in the youth population coincides with a number of changes in the human capital endowments of youth that will most likely constrain military recruitment objectives in the future. The target population seems to be split by an educational divide, which renders recruitment more difficult in quantitative and qualitative terms. Until 2030, the absolute number of graduates from lower and middle level institutions of education is projected to decline, just as the number of youth leaving school without a certificate. At the same time, the number of graduates with upper level education is projected to increase. Following one of the main interests of this study, several projections of youth's future human capital endowments were calculated. Accordingly, the contraction of the traditional target age group falls together with an expansion of adverse health factors that may disqualify an increasing share of youth from military service, in particular at the lower and middle educational level. The group of less educated males and females, that commonly shows a relatively large propensity for military service, will decline in absolute terms and feature a disproportionate reduction of persons without any potentially disqualifying heath conditions. In sum, the military recruitment potential will not only be reduced in quantitative terms. A higher prevalence of disqualifying attributes, affecting males and females at all educational levels, will additionally limit the military recruitment potential. In view of these unfavorable qualitative trends, there is indication that the real recruitment challenge in the future will be more severe than exclusively quantitative projections would suggest. However, policies can affect many of the relevant variables.

7.2 Policy Options

The Bundeswehr is not alone in facing recruitment challenges. Particularly in the transition from a conscripted to an all-volunteer force, allied partners experienced severe difficulties in attracting recruits of sufficient quantity and quality. In view of the suspension of conscription in Germany, there is an increasing importance of effective military recruitment. Thus far, the manpower demand of the Bundeswehr was, at least, partially met by the inflow of conscripts. Between 30 and 40 % of the military personnel initially passed the more or less obligatory military service. This reliable source of military recruitment has now run dry.

In an international comparison of approaches to military recruitment, it appears that some armed forces are more innovative than others. Likewise noticeable is the fact that the armed forces' degree of preparedness does not correspond to the magnitude of demographic change. That means that the armed forces of countries with a relatively favorable or, at least, stable demographic outlook like Belgium, Canada, the Netherlands, the United Kingdom or the United States are relatively well-positioned for the fiercer competition from private and public organizations for suitable personnel. The Bundeswehr faces an unequally steeper decline of the youth population than any other Western partner (see Fig. 7.1). Albeit existing recruitment strategies, even those under development within the current Bundeswehr reform, may not go far enough to

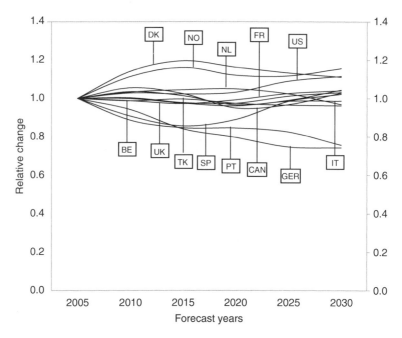

Fig. 7.1 Relative change in youth by country (Source: UN 2010, medium variant, own estimations. *BE* Belgium, *CAN* Canada, *DK* Denmark, *FR* France, *GER* Germany, *IT* Italy, *NL* Netherlands, *NO* Norway, *PT* Portugal, *SP* Spain, *TK* Turkey, *UK* United Kingdom, *US* United States of America)

compensate for the lost "personnel reservoir" of conscripts and secure a continuous inflow of enough new soldiers into the organization.

There is wide variety in the political, economic and social conditions among member states of NATO or the European Union. Still, it is possible to draw some general conclusions about critical success factors in recruiting policies and necessary tradeoffs between objectives for quantity and quality. In general, however, allied military forces faced significant recruitment challenges transitioning to an all-volunteer force with costs for effective recruiting measures often exceeding the budget plan. In particular, measures designed to make a military career more attractive and improve the working conditions for soldiers turned out to be overly expensive. These costs necessary to retain the military as a viable employer will keep rising in an increasingly competitive labor market.

With the private sector offering better financial and non-financial recruitment incentives, Western militaries already experience personnel shortages in critical functions. Particularly, technical and medical positions are hard to fill. The same is true for jobs with higher-than-average requirements in physical fitness, such as in regular infantry or the special forces. In terms of ranks, the recruitment of non-commissioned officers poses the most significant challenge (Moelker et al. 2005: 23f.; Szvircsev Tresch 2010: 147ff.). In the long-run, the inability to fill vacant positions or attract qualified personnel may have operational consequences and hamper military effectiveness. For example, the deployment of Dutch troops in Eritrea/Ethiopia and Kosovo was limited to short periods of time due to manning problems (Moelker et al. 2005: 25). Hence, attracting enough suitable recruits has become a key strategic element.

In a survey of the recruiting efforts and results of allied forces, several factors of success and categories of recruitment measures can be identified. In order to compensate for the suspension of conscription and maintain a similar level of familiarity with the armed forces, several partners extended pre-recruitment measures and established new points of contacts with young people. Wherever quantitative effects were considered a priority, new applicant pools were tapped by altering the existing recruitment criteria of age, quality and nationality. Other measures targeted the basic conditions of employment in terms of pay, contract length, work-life balance and military facilities. Quick-win measures improved the efficiency of the recruitment process and thereby reduced the number of dropouts during the time between application, selection and employment. Internal communication campaigns raised awareness for the recruitment shortfalls, whereas external marketing efforts portrayed the military as an employer with interesting jobs and opportunities for professional growth. The terms of the specific package of measures varied due to differences in the availability of financial means and priorities in terms of recruits' quantity and quality. The following overview of recruitment policies adopted by allied forces may provide the Bundeswehr with some insight for its transition to the all-volunteer force and ever tougher competition for talents from civilian workforce employers. Many of the measures have the potential to make the Bundeswehr less vulnerable to the imminent demographic reduction in its primary target population.

7.2.1 Pre-recruitment Measures

Research on military propensity among young people has shown that interest in a military career declines with age (Bulmahn et al. 2010: 9; van de Ven and Bergman 2007: 2D-4). Therefore, there has been a growing consensus that recruiting initiatives should not only inform but already commit youths well before they reach the common service age of 18 years (Szvircsev Tresch 2010: 158). In many countries, however, adolescents under the age of 18 are not allowed to serve in the military as regular soldiers. Several allied partners have established pre-recruitment arrangements to attract and bind young prospects to the military (Moelker et al. 2005: 25). For example, the Dutch military offers an "orientation year", which comprises a civilian upper secondary vocational education at regional training centers. About 20 % of the curriculum consists in physical exercise, military-purpose education and information about the armed forces. Participants below the age 18 have the status of an "aspirant soldier". They do not hold a regular position and are not allowed to carry weapons or be sent on deployments (van de Ven and Bergman 2007: 2D-4).

According to official sources, the "orientation year" of the Dutch military proved highly effective. Drop-out rates were about 10 %, and more than 60 % of those initially starting the program enlisted in the military afterwards. Newly enlisted soldiers, having undergone this orientation period, have more accurate expectations about military service. Early turnover during the regular military training is much lower among them compared to those enlistees without preparatory training (van de Ven and Bergman 2007: 2D-4).

For high-school graduates at the age of 16, the Dutch military additionally offers a 2-year program in a so-called "Peace and Security School" that provides further education in security-related issues. Just as the 1-year orientation program, such military schools allow young people to bridge the time gap between graduation and becoming regular soldiers. The "Peace and Security School" is an equally effective source of military recruitment. On average, about 2,000 adolescents participate in the program, and more than half of them eventually join the Dutch military (Szvircsev Tresch 2010: 152). Similar pre-recruitment institutions exist in Belgium, Turkey and the United Kingdom (NATO 2007).

On a lower level of commitment, young people in the Netherlands could attend so-called "familiarization weeks" and introduced to the military profession through bivouacking, information sessions, and a variety of typical activities. The objective of the program was to positively influence youth's interest in a military career, increase their intention to apply, and reduce the risk of early attrition due to unrealistic expectations (van de Ven and Bergman 2007: 2D-3). After a pilot project in 1998, 8 of these "familiarization weeks" were organized in 1999, in which 362 adolescents participated. Despite the positive feedback, the program was suspended due to the extra workload for the training centers (Williams and Seibert 2011: 21). The Belgian military pursued a similar approach and objective when organizing youth camps for adolescents of about 16 years (Lescreve and Schreurs 2007: 2A-8).

7.2.2 Expansion of the Recruitment Potential

To alleviate shortfalls in recruitment, several allied forces expanded the recruitable population along criteria like age, gender, quality and citizenship. With regard to age, in many places, the target audience was extended to include those under the age of 18 in the framework of pre-recruitment measures (see section before) and those above the original maximum age around the age of 30. For example, the British military focus its recruiting efforts on the 16-year-old age group and increased the maximum age at which soldiers can enlist from 26 to 33 years (Dandeker and Mason 2010: 217). Similarly, in Belgium, the maximum age for applicants was raised from 31 to 34 years. Medical doctors are eligible for lateral entry into the Belgian military until they are 49 years given that they have professional experience of at least 5 years (Lescreve and Schreurs 2007: 2A-9).

Another possibility is to change the qualitative selection parameters by lowering the cut-off scores of military entry tests or easing the minimum schooling requirements.

For example, the Belgian and Spanish military found this to be an efficient way to fill open vacancies (NATO 2007). However, if the achievement of quantitative objectives stands above the fulfillment of qualitative criteria, the proclaimed goal of recruiting and developing a "quality soldier" becomes unfounded (Frieyro de Lara 2010: 182). For this reason, several allied forces adhere to their entry standards even though that means that they cannot reach their recruiting targets (Szvircsev Tresch 2010: 158). The main reason is two-fold: cost avoidance and lower attrition. When selection levels are low, the duration of military education and training has to be extended (Moelker et al. 2005: 25). Against this background, the Canadian military implemented a shift from recruiting mostly unskilled applicants to attracting skilled and experienced applicants that can skip as much training as possible and fill vacancies quickly (Syed and Morrow 2007: 2B-4). In addition, there is empirical evidence that better educated recruits have a lower risk of early attrition. One consequence of the Belgian approach of "filling all vacancies" is a high turnover rate during the first 2 years of basic training (Szvircsev Tresch 2010: 158). When newly recruited and trained soldiers quit service prematurely, it is very costly for the military since there is no return on investment.

Other applicants are motivated and cognitively capable but fail the fitness test. The Dutch military offers those rejected applicants a personalized training recommendation to improve their physical strength and stamina. For 3 months, they can work out in a gym at the expense of the defense organization before retaking the fitness test. The program proved highly successful. About half of the initially rejected applicants returned and passed the fitness test (van de Ven and Bergman 2007: 2D-4).

Another strategy for broadening the recruitment pool is to target segments of the population that are thus far underrepresented in the armed forces, such as women and members of minority ethnic communities. In Germany, the Bundeswehr allows women to serve in the military since 2001. Today about 17,580 soldiers or 9.3 % of all soldiers in the Bundeswehr the armed forces are women (German Bundestag 2011: 24). In other countries, the share of female soldiers is higher. For example, in Canada, France and Spain the share of female personnel ranged between 12 and 17 % in the 2007–2008 period (Schjølset 2010: 28). In Canada, the so-called Employment Equity Act requires the military to reflect the demographic composition of the civilian workforce in its personnel structure. Recruiting initiatives therefore target explicitly women and visible minority groups (Syed and Morrow 2007: 2B-2). The Belgian military introduced gender-specific cut-off scores for the physical fitness test in order to prevent female applicants to fail standards that appeared unwarrantedly high (Lescreve and Schreurs 2007: 2A-4). In the Netherlands, the 2004 Plan of Action for Gender Issues of the Dutch military mandated an intake of 30 % of females (van de Ven and Bergman 2007: 2D-12).

Besides women, immigrants seem to have become the solution for recruitment problems among many allied partners (Frieyro de Lara 2010: 190). Such a diversification of recruitment pools secures access to sections of the population that have been significantly under-represented in the military and that constitute a relatively large share of the younger age groups (Dandeker and Mason 2010: 218). Hence, the British military enlists soldiers from other Commonwealth states since 1998

(Ware 2010: 324). In that year, the national Strategic Defense and Security Review mandated that the British military should reflect the ethnic composition of the general society (Mason and Dandeker 2009: 394). This objective was achieved in April 2009, when about 10,500 soldiers from 38 nations served in the British military, corresponding to a share of about 11 %. The French military collects no data on ethnic minorities officially. However, observers estimate the share of ethnic minority soldiers ranges between 10 and 20 % (Williams and Seibert 2011: 13f.). The Spanish military targets foreigners living in Spain, in particular those originating from Latin America and Equatorial Guinea. According to current legislation, there is a 9 % ceiling on foreign soldiers, and they already make up 7.3 % of the armed forces personnel. A provision was implemented that enables soldiers of foreign nationality to obtain Spanish citizenship (Frieyro de Lara 2010: 186). In the Netherlands, the recruitment targets for ethnic minority soldiers are set in accordance with the proportion of ethnic cultural minorities in the total Dutch workforce. In 2006, this target value was 7.6 % (Richardson 2010: 28).

However, there are two main objections raised against increasing the minority share in the armed forces: worsening of cohesion and lack of interest among the actual target group. Yet there seems no empirical evidence that diversity undermines cohesion. Rather what counts are qualitative attributes like skills, knowledge, abilities, and attitudes (Leuprecht 2010: 39ff.). However, research on military propensity among minority groups confirms their relatively meager interest in a career with the armed forces. In the Netherlands, surveyed youth of Turkish, Moroccan, Suriname or Arabian descent cite the weak "diversity profile" of the military and specific job characteristics as the main factors that impinge on their interest in a military career. At the same time, the military faces considerable competition in minority recruitment from other government bodies, especially from the police (Richardson 2010: 28). Similarly, the British military is not seen as an attractive career choice among second-generation immigrants in the United Kingdom. According to a survey among youth of Pakistani descent, higher education is deemed more desirable than a military career (Mason and Dandeker 2009: 401). Hence, the local recruitment potential among ethnic minorities is bounded, and those sections of the population with a younger age profile hard to reach for the military.

Of all policy options, the expansion of the recruitable population would have the most significant direct effect on military manpower supply. By making entry requirements relating to age, sex, nationality and qualifications more flexible, the Bundeswehr would be able to recruit thus far underrepresented segments of the population (e.g. females, as well as ethnocultural and religious minorities with German citizenship) and enlist previously excluded groups of people. This would not only enhance the representation of the population at large; it would also be in line with economic and strategic rationale. Just as in the civilian economy, females and previously disadvantaged minorities would become the major beneficiaries of demographic change in the military organization (Weiner and Russell 2001: 136). An increasing heterogeneity of the military workforce would also have a direct influence on the strategic options of the Bundeswehr. During military operations abroad, ethnocultural diversity among deployed soldiers may serve as "multiplier" of capabilities that increases the chances of mission success in culturally complex settings (Leuprecht 2006: 33). During the operations in Afghanistan, German soldiers with Turkish roots were characterized by a heightened cultural sensitivity, while female soldiers deployed to military operations in Afghanistan and Kosovo were particularly skillful in establishing interactions with the local population (van der Meulen and Soeters 2007).

7.2.3 Modification of Length and Type of Service

In response to social change and a new social interpretation of gainful employment, most armed forces changed the terms of enlistment and now offer different forms of military service with varying durations of service (Szvircsev Tresch 2010: 160). For example, the Belgian military introduced short-term contracts of 2 years to maintain an acceptable age structure. Soldiers can renew their contract for a total of 7 years until they reach the age of 34 (Lescreve and Schreurs 2007: 2A-4). The French military offers more flexible terms of service to better suit the life circumstances of recruits. Short-term or intermittent military duty are depicted as a bridging period in the personal life cycle (Frisch 2011: 6). The U.S. military raised its maximum age for enlistment from 35 to 42 years and reduced the enlistment period for some recruits from 4 years to 15 months (Korb and Duggan 2007: 468). In contrast, the British military considers it more advantageous to retain experienced personnel than recruiting and training short-term personnel on a continuous basis (Williams and Seibert 2011: 16f.).

7.2.4 Increase of Pay and Benefits

Material benefits are an essential component in attracting youth to enter the armed forces. This is all the more true for all-volunteer forces, for which the marketplace plays a fundamental role (Korb and Duggan 2007: 468). In the British military, general annual pay raises are linked to the increase in civilian wages. An additional payment, known as the X-factor, compensates for the disadvantages of service life. In 2001, the X-factor was set at 13 %, which means that earnings in the armed forces should be 13 % higher than earnings in a comparable job in the civilian sector (Hartley 2006: 308). In addition to these across-the-board pay raises for all military personnel, the British military launched financial incentives for selective vacancies that were hard to fill (Frisch 2011: 7). The U.S. military offers much larger cash bonuses and incentives to attract new recruits. For example, the army offered up to 15,000 for enlisted job categories, up to 20,000 dollars for recruits that score well on entrance tests, and up to 40,000 dollars for specialty assignments with advanced linguistic and specific civilian skill requirements (Korb and Duggan 2007: 468; Moniz 2005).

7.2.5   More Efficient Recruitment Processes

To make their recruitment processes more efficient and reduce the attrition of applicants during the selection process, several armed forces have adopted mix of measures in the realm of data management and communications. The Dutch military reduced its response time on applications and selection procedures. About 82 % of the applications are processed within ten working days, after starting their cognitive, psychological, and medical examination (Richardson 2010: 28). In order to inform applicants as soon as possible, the Spanish military implemented a computerized recruiting management system in 2003. The system connects in real time the various recruitment directorates with the local recruitment centers. Simultaneously, the Spanish military introduced penalties for applicants that drop out shortly before or after the beginning of the military training. Those individuals are not allowed to reapply for the next five application cycles (Puente and Blanco 2007: 2E-10). The Canadian military improved the process of recruitment by introducing an e-recruiting platform in 2005. It consists of an on-line application service and also enables recruiters to quickly establish contacts with applicants. At the same time, the Canadian forces launched an initiate to replace paper and pencil in the recruitment process, for example by means of virtual medical files, electronic aptitude tests or applicant surveys (Syed and Morrow 2007: 2B-6).

7.2.6   More Effective Advertising and Communication

An integral part of a successful recruitment strategy is to provide prospective applicants and their social environments with accurate information about jobs in the armed forces. By communicating a constant theme, the military must be recognized as a reputable employer, and all vocational opportunities must be known (Szvircsev Tresch 2010: 160). Misconceptions about entry requirements, ignorance of available career options, uncertainty about the military training program, concerns about military discipline or the compatibility of family and career may act as significant barriers to enlistment (Johansen 2007: 2G-4). Common recruitment campaigns highlight the opportunities for both professional and personal development within short-term or long-term military careers, competitive salaries, important aspects of quality of life, financial incentives such as recruiting allowances, and other potential benefits. In this connection, the British military acts on the maxim: "People need to know what they do, how it is done, what the values and rewards are" (Johansen 2007: 2G-4).

Current military service personnel may act as a trusted and reliable source of information for potential recruits. Several armed forces therefore enhanced their internal communication to service members. The Canadian forces offer information through internal publications or staff briefings to keep everybody informed about ongoing recruiting initiatives. An additional communications campaign "Recruiting is everybody's business" addresses all military personnel to engage in the comprehensive recruitment strategy and connect with the broader population (Syed and Morrow 2007: 2B-6). Similarly, the British military used the slogan "it takes a soldier to recruit a soldier" and increasingly sent young, recently trained soldiers to their hometowns and schools to talk about their military experience with peers (Johansen 2007: 2G-4). For every new recruit, existing soldiers receive a bonus of currently 1.300 lb from the British military (Dandeker and Mason 2010: 217). On this basis, the Dutch military introduced a similar bonus scheme that awards a bonus of 1.000 Euros to military personnel that introduce a new recruit (Richardson 2010: 28).

One of the downsides of the "selling approach" in military recruitment are the high rates of attrition during the first few months of service. For example, about 30 % of the new recruits in the Belgian military voluntarily withdraw from initial training. Presumably, the major reason consists in the idealization of military service in contemporary recruitment campaigns and the fact that only positive aspects are communicated to prospective applicants (Lescreve and Schreurs 2007: 2A-6). Besides inflated expectations, one of the reported reasons for early turnover in the Belgian military was the tough stance of the drill instructors towards new recruits. Hence, an experiment showed that attrition rates were much lower when only older instructors held the initial training program (Lescreve and Schreurs 2007: 2A-8). Such a solution would also be in line with the demographic changes ahead.

7.3 Conclusion

This book contributes to a growing literature about the security implications of demographic change. It rests on the assumption that contemporary military interventions still place great emphasis on manpower. In view of the strategic importance to operational readiness, the book provides a detailed analysis of the impact of demographic change on military recruitment in Germany.

Since the military recruitment potential is proportional to the size of the youth population, changes therein may constrain the intervention capability of the Bundeswehr and the foreign policy capacity of Germany overall. However, there is a range of qualifying parameters needs that needs to be taken into account when considering demographic change as an indicator of military recruitment in the future. One is education. While it is true that there has been a universal trend towards higher education, the youth population at recruitable age in Germany seems to be split by an educational divide: A large part pursues tertiary education, whereas another significant share does not fulfill the cognitive entry requirements of the Bundeswehr. A second mitigating factor is health. Soldiers' physical and mental requirements have been on a continuous rise. Yet, recent trends indicate substantial health risks and deficits among youth in Germany, e.g. overweight and obesity, physical inactivity, or smoking. Third is military propensity. The Bundeswehr, just like any other military that depends on volunteers, can only recruit as many soldiers as want to join the military. However, it can be assumed that in the foreseeable future, the attractiveness of the soldier profession, relative to other civilian employment, will further decline. On the one hand, civilian institutions will launch attractive recruitment incentives in light of the decline in the population of working age. On the other, a military career is characterized by various hardships that may be perceived increasingly unappealing, including long absences from home, high demands on the geographic mobility of soldiers and their families, risks to life and limb, and a high degree of bureaucracy and rigidity.

Although the numbers of those willing to serve can be altered by recruitment policies and financial incentives, spending less rather than more on personnel is essential to have sufficient funds available for state-of-the-art military technology that secures the operational readiness of the Bundeswehr and also minimizes casualties. In part, these financial constraints result from the fiscal and political consequences of demographic change. In view of the increased public spending on the elderly population and their preference for domestic (over foreign) policy priorities, maintaining military expenditures at the current level will, in all likelihood, become more difficult.

This interaction of a shrinking military recruitment potential and cuts to military spending may lead to a situation of "structural disarmament", where the Bundeswehr can no longer be considered capable of manpower-intensive operations, such as counterinsurgency or military operations other than war in failing or failed states, and Germany will have to reduce its military and related contributions to international security accordingly, thereby taking a strong moral, political and diplomatic stance. As Germany will not be alone in tailoring its political-military strategy towards the new demographic and economic realities, multilateral security cooperation among countries with stagnant and aging or even shrinking populations will become increasingly important for coping with international security threats, providing effective military capabilities, and maintaining international credibility. Targeting unexploited recruitment potentials can only be a temporary, though partial and not entirely unproblematic, relief for the challenges ahead of the Bundeswehr.

Bibliography

Agakimi, H. (2006). We the Japanese people. A reflection on public opinion. Tokyo: The Japan Institute of International Affairs. Commentary, online at http://yaleglobal.yale.edu/display. article?id=7444. Accessed 10 Aug 2009.

Agresti, A., & Finlay, B. (1999). *Statistical methods for the social sciences*. Upper Saddle River: Prentice-Hall.

AIK. (2006). Jugendumfrage. Meinungsbild der Jugend zur sicherheitspolitischen Lage in der Bundesrepublik Deutschland. Scientific use file. Bielefeld/Strausberg: TNS Emnid/Akademie der Bundeswehr für Information und Kommunikation (AIK).

Aldis, A., & Herd, G. (2004). Managing soft security threats: Current progress and future prospects. *European Security, 13*(1), 169–186.

Alexander, M., & Garden, T. (2001). The arithmetic of defense policy. *International Affairs, 77*(3), 509–529.

Apfelbacher, C. J., Cairns, J., Bruckner, T., Mohrenschlager, M., Behrendt, H., Ring, J., et al. (2008). Prevalence of overweight and obesity in East and West German children in the decade after reunification: Population-based series of cross-sectional studies. *Journal of Epidemiology and Community Health, 62*(2), 125–130. doi:10.1136/jech.2007.062117.

Apt, W. (2011). Aufstand der Jugend: Demographie liefert Hinweise auf Konfliktpotentiale. *SWP-Aktuell, A, 16*, 4.

Artelt, C., Baumert, J., Klieme, E., Neubrand, M., Prenzel, M., Schiefele, U., et al. (2001a). *Pisa 2000: Zusammenfassung Zentrale Befunde. Schülerleistungen im internationalen Vergleich* (p. 51). Berlin: Max-Planck-Institut für Bildungsforschung.

Artelt, C., Stanat, P., Schneider, W., & Schiefele, U. (2001b). Lesekompetenz: Testkonzeption und Ergebnisse. In J. Baumert, E. Klieme, M. Neubrand, M. Prenzel, U. Schiefele, W. Schneider, et al. (Eds.), *Pisa 2000: Basiskompetenzen von Schülerinnen und Schülern im internationalen Vergleich* (pp. 69–137). Opladen: Leske + Budrich.

Asch, B. J., Kilburn, M. R., & Klerman, J. A. (1999). *Attracting college-bound youth into the military. Toward the development of new recruiting policy options* (p. 61). Santa Monica: RAND Corporation.

Asch, B. J., Romley, J., & Totten, M. (2005). *The quality of personnel in the enlisted ranks* (p. 102). Santa Monica: RAND.

Bachman, J. G., Segal, D. R., Freedman-Doan, P., & O'Malley, P. M. (1998). *Military propensity and enlistment: Cross-sectional and panel analyses of correlates and predictors* (Monitoring the future occasional paper 41, p. 233). Ann Arbor: Michigan University.

Bachman, J. G., Segal, D. R., Freedman-Doan, P., & O'Malley, P. M. (2000). Who chooses military service? Correlates of propensity and enlistment in the U.S. armed forces. *Military Psychology, 12*(1), 1–30.

W. Apt, *Germany's New Security Demographics: Military Recruitment in the Era of Population Aging*, Demographic Research Monographs, DOI 10.1007/978-94-007-6964-9, © Springer Science+Business Media Dordrecht 2014

Backhaus, K., Erichson, B., Plinke, W., & Weiber, R. (2000). *Multivariate analysenmethoden*. Berlin: Springer.

Battistelli, F. (1997). Peacekeeping and the postmodern soldier. *Armed Forces & Society, 23*(3), 467–484.

Beck, H.-C. (1998). Innere Führung heute. In *Öffentliche Gelöbnisse und Innere Führung, Wolf Graf von Baudissin-Memorial-Symposium: Vol. Heft 112* (pp. 20–46). Hamburg: Institut für Friedensforschung und Sicherheitspolitik an der Universität Hamburg (IFSH).

Becker, G. S. (1962). Investment in human capital: A theoretical analysis. *Journal of Political Economy, 70*(5), 9–49.

Becker, G. S. (1964). *Human capital: A theoretical and empirical analysis, with special reference to education* (p. 187). New York/London: National Bureau of Economic Research.

Becker, G. S. (1975). *Human capital: A theoretical and empirical analysis, with special reference to education* (2nd ed.). New York: Columbia University Press.

Becker, G. S. (1981). *A treatise on the family*. Cambridge, MA: Harvard University Press.

Ben-Ari, E. (2005). Epilogue: A "good" military death. *Armed Forces & Society, 31*(4), 651–664.

Bernard, M., & Phillips, J. (2000). The challenge of ageing in tomorrow's Britain. *Ageing and Society, 20*, 33–54.

Berner, J. K., & Daula, T. V. (1993). *Recruiting goals, regime shifts, and the supply of labor to the army* (p. 43). West Point/New York: USMA Operations Research Center.

Berry, W. D., & Lowery, D. (1990). An alternative approach to understanding budgetary trade-offs. *American International Studies Quarterly, 42*, 645–674.

Bicksler, B. A., & Nolan, L. G. (2006). Recruiting an all-volunteer force: The need for sustained investment in recruiting resources. *Policy Perspectives, 1*(1), 32.

Biddle, S. (2002). *Afghanistan and the future of warfare: Implications for army and defense policy* (p. 68). Carlisle: Strategic Studies Institute, United States Army War College.

Biehl, H., & Kümmel, G. (2003). Military service. In J. Callaghan & F. Kernic (Eds.), *Armed forces and international society. Global trends and issues*. Münster: LIT Verlag.

Biehl, H., Hagen, U. v., & Mackewitsch, R. (2000). *Motivation von Soldaten im Auslandseinsatz* (SOWI-arbeitspapier Nr. 125, p. 55). Strausberg: Sozialwissenschaftliches Institut der Bundeswehr.

Biehl, H., Klein, P., & Kümmel, G. (2007). Diversity in the German armed forces. In J. Soeters & J. v. d. Meulen (Eds.), *Cultural diversity in the armed forces. An international comparison* (pp. 171–184). Oxon: Routledge.

Biermann, F., Petschel-Held, G., & Rohloff, C. (1998). Umweltzerstörung als Konfliktursache? Theoretische Konzeptualisierung und empirische Analyse des Zusammenhangs von "Umwelt" und "Sicherheit". *Zeitschrift für Internationale Beziehungen, 5*(2), S.273– S.308.

Binkin, M. (1986). *Military technology and defense manpower* (Studies in defense policy, p. 144). Washington, DC: Brookings Institution Press.

Boehmer, M., Zucker, A., Ebarvia, B., Seghers, R., & Snyder, D. (2003). June 2003 youth poll: Overview report (pp. 157). Arlington: U.S. Department of Defense, Department of Joint Advertising, Market Research and Studies.

Boëne, B. (2003). The military as a tribe among tribes. Postmodern armed forces and civil-military relations? In G. Caforio (Ed.), *Handbook of the sociology of the military* (pp. 167–187). New York: Kluwer Academic/Plenum Publishers.

Bonin, H., Schneider, M., Quinke, H., Arens, T. (2007). *Zukunft von Bildung und Arbeit. Perspektiven von Arbeitskräftebedarf und -angebot bis 2020* (IZA Research Report No. 9, p. 213). Bonn

Boot, M. (2005). The struggle to transform the military. *Foreign Affairs, 84*(2), 103–118.

Booth, H. (2006). Demographic forecasting: 1980 to 2005 in review. *International Journal of Forecasting, 22*(3), 547–581.

Brennan, E. (1999). *Population, urbanization, environment, and security: A summary of the issues* (Comparative urban studies occasional series, 22). Washington, DC: Woodrow Wilson International Center for Scholars.

Brooks, R. (2000). *Population aging and global capital flows in a parallel universe* (p. 31). Washington, DC: International Monetary Fund.

Brown, M. E. (2003). Introduction: Security challenges in the twenty-first century. In M. E. Brown (Ed.), *Grave new world: Security challenges in the 21st century* (pp. 1–16). Washington, DC: Georgetown University Press.

Bulmahn, T. (2007). *Berufswahl Jugendlicher und Interesse an einer Berufstätigkeit bei der Bundeswehr: Ergebnisse der Jugendstudie 2006 des Sozialwissenschaftliches Instituts der Bundeswehr* (p. 107). Strausberg: Sozialwissenschaftliches Institut der Bundeswehr.

Bulmahn, T. (2008). *Bevölkerungsbefragung 2008. Sicherheits- und verteidigungspolitisches Meinungsklima in Deutschland. Kurzbericht* (p. 46). Strausberg: Sozialwissenschaftliches Institut der Bundeswehr.

Bulmahn, T., Fiebig, R., Greif, S., Jonas, A., Sender, W., & Wieninger, V. (2008). *Sicherheits- und verteidigungspolitisches Meinungsklima in der Bundesrepublik Deutschland: Ergebnisse der Bevölkerungsbefragung 2007* (p. 201). Strausberg: Sozialwissenschaftliches Institut der Bundeswehr.

Bulmahn, T., Fiebig, R., Hennig, J., & Wieninger, V. (2010). *Ergebnisse der Jugendstudie 2008* (p. 156). Strausberg: Sozialwissenschaftliches Institut der Bundeswehr.

Burk, J. (1994). Thinking through the end of the cold war. In J. Burk (Ed.), *The military in new times: Adapting armed forces to a turbulent world* (pp. 1–24). Boulder: Westview Press.

Burk, J. S., & Faris, J. H. (1982). *The persistence and importance of patriotism in the all-volunteer forces*. Report to the U.S. Army Recruiting Command (p. 124). Washington, DC: Batelle Memorial Institute.

Buse, U. (2011, May 30). Die Drückerkompanie. *DER SPIEGEL*, pp. 50–54.

Buzan, B. (1997). Rethinking security after the cold war. *Cooperation and Conflict, 32*(5), 5–28.

Buzan, B., Waever, O., & de Wilde, J. (1993). *Security: A new framework for analysis*. London: Lynne Rienner Publishers.

Cain, M. (1983). Fertility as an adjustment to risk. *Population and Development Review, 9*(4), 688–702.

Cain, G. G. (1986). The economic analysis of labor market discrimination: A survey. In O. Ashenfelter & R. Layard (Eds.), *Handbook of labor economics* (Vol. I, pp. 693–785). Amsterdam: North Holland.

Carlsnaes, W. (2006). Foreign policy. In W. Carlsnaes, B. A. Simmons, & T. Risse (Eds.), *Handbook of international relations* (pp. 331–349). London/Thousand Oaks/New Delhi: Sage Publications.

Carr, P., & Kefalas, M. J. (2009). *Hollowing out the middle: The rural brain drain and what it means for America*. Boston: Beacon.

Carr, D. L., Markusen, J. R., & Maskus, K. E. (2007). Investment in developing countries: Human capital, infrastructure, and market size. In R. E. Baldwin & L. A. Winters (Eds.), *Challenges to globalization: Analyzing the economics* (pp. 383–409). Chicago: University of Chicago Press.

Caselli, G., Meslé, F., & Vallin, F. (2002). Epidemiologic transition theory exceptions. *Genus, 9*, 9–51.

Cedefop. (2008). *Future skill needs in Europe: Medium-term forecast*. Synthesis report (pp. 129). Luxembourg: European Centre for the Development of Vocational Training (Cedefop), Office for Official Publications of the European Communities.

Chan, S. (1995). Grasping the peace dividend: Some propositions on the conversion of swords into plowshares. *Mershon International Studies Review, 39*(1), 53–95.

Choucri, N. (1974). *Population dynamics and international violence*. Lexington: Heath.

Choucri, N., & North, R. C. (1972). Dynamics of international conflict: Some policy implications of population, resources, and technology. *World Politics, 24*, 80–122. Supplement: Theory and Policy in International Relations.

CIA. (2001). *Long-term global demographic trends: Reshaping the geopolitical landscape*. Washington, DC: Central Intelligence Agency.

Cincotta, R. P. (2004). Demographic security comes of age. *Environmental Change and Security Project Report, 10*, 24–29.

Cincotta, R. P., Engelman, R., & Anastasion, D. (2003). *The security demographic: Population and civil conflict after the cold war*. Washington, DC: Population Action International.

Clark, D. H. (2001). Trading butter for guns: Domestic imperatives for foreign policy substitution. *Journal of Conflict Resolution, 45*(5), 636–660.

Clark, D., & Hart, R. (2003). Paying the piper? Implications of social insurance payments for conflict propensity. *International Interactions, 29*(1), 57–82.

Cohen, E. A. (1996). A revolution in warfare. *Foreign Affairs, 75*(2), 37–54.

Cohn, L. (2007). *Who will serve? Education, labor markets, and military personnel policy.* Durham: Department of Political Science, Duke University.

Coleman, D. (2006). Immigration and ethnic change in low-fertility countries: A third demographic transition. *Population and Development Review, 32*(3), 401–446.

Coleman, D. (2012). The changing face of Europe. In J. A. Goldstone, E. P. Kaufmann, & M. D. Toft (Eds.), *Political demography. How population changes are reshaping international security and national politics* (pp. 176–193). New York: Oxford University Press. 342.

Collier, P., & Hoeffler, A. (2004). Greed and grievance in civil war. *Oxford Economic Papers, 56*(4), 563–595.

Cote, O. R. (2004). The personnel needs of the future force. In C. Williams (Ed.), *Filling the ranks: Transforming the U.S. military personnel system* (Belfer Center for Science and International Affairs, pp. 55–68). Cambridge, MA: John F. Kennedy School of Government, Harvard University.

Dale, C., & Gilroy, C. (1983). *The economic determinants of military enlistment rates* (p. 57). Alexandria: U.S. Army Research Institute for the Behavioral and Social Sciences.

Dalgaard-Nielsen, A. (2006). *Germany, pacifism, and peace enforcement.* Manchester/New York: Manchester University Press.

Dalton, R. J. (1988). *Citizen politics in western democracies: Public opinion and political parties in the United States, Great Britain, West Germany, and France* (270pp.). Chatham: Chatham House Publishers.

Dandeker, C. (1994). A Farwell to arms? The military and the nation-state in a changing world. In J. Burk (Ed.), *The military in new times: Adapting armed forces to a turbulent world* (pp. 117–139). Boulder: Westview Press.

Dandeker, C. (1995). Flexible forces for a post-cold war world: A view from the United Kingdom. *The Tocqueville Review, XVII*(1), 23–38.

Dandeker, C. (2003). Building flexible forces for the 21st century. Key challenges for the contemporary armed services. In G. Caforio (Ed.), *Handbook of the sociology of the military* (pp. 405–416). New York: Kluwer Academic/Plenum Publishers.

Dandeker, C., & Mason, D. (2010). Echoes of empire: Addressing gaps in recruitment and retention in the British army by diversifying recruitment pools. In T. Szvircsev Tresch & C. Leuprecht (Eds.), *Europe without soldiers? Recruitment and retention across the armed forces of Europe* (Queen's policy studies series, pp. 209–231). Montreal/Kingston: McGill-Queen's University Press.

Daniels, S. R. (2006). Critical periods for abnormal weight gain in children and adolescents. In M. I. Goran & M. S. Sothern (Eds.), *Handbook of pediatric obesity: Etiology, pathophysiology, and prevention* (pp. 67–78). Boca Raton: CRC Press.

Däniker, G. (1992). *Wende Golfkrieg: Vom Wesen und Gebrauch künftiger Streitkräfte.* Report Verlag: Frankfurt/Main.

Demeny, P., & McNicoll, G. (2006). The political demography of the world system, 2000–2050. In P. Demeny & G. McNicoll (Eds.), *The political economy of global population change, 1950–2050* (Population and development review: Supplement 32, pp. 254–287). New York: The Population Council.

Dertouzos, J. N. (1985). *Recruiter incentives and enlistment supply* (p. 55). Santa Monica: RAND Corporation.

Dhillon, N., & Yousef, T. (2009). *Generation in waiting: The unfulfilled promise of young people in the Middle East.* Washington, DC: Brookings Institution Press.

Domke, W. K., Eichenberg, R. C., & Kelleher, C. M. (1983). The illusion of choice: Defense and welfare in advanced industrial democracies, 1948–1978. *The American Political Science Review, 77*(1), 19–35.

Dorbritz, J., & Ruckdeschel, K. (2007). Kinderlosigkeit in Deutschland: Ein europäischer Sonderweg? In D. Konietzka & M. Kreyenfeld (Eds.), *Ein Leben ohne Kinder. Kinderlosigkeit in Deutschland* (pp. 45–81). Wiesbaden: VS Verlag.

Dorbritz, J., & Schwarz, K. (1996). Kinderlosigkeit in Deutschland: Ein Massenphänomen? Analysen zu Erscheinungsformen und Ursachen. *Zeitschrift für Bevölkerungswissenschaft, 3*, 231–261.

Dorbritz, J., Ette, A., Gärtner, K., Grünheid, E., Mai, R., Micheel, F., et al. (2008). In Bundesinstitut für Bevölkerungsforschung in Zusammenarbeit mit dem Statistischen Bundesamt (Ed.), *Bevölkerung: Daten, Fakten, Trends zum demographischen Wandel in Deutschland* (pp. 80). Wiesbaden.

Easterlin, R. A. (1978). What will 1984 be like? Socioeconomic implications of recent twists in age structure. *Demography, 15*(4), 397–432.

Ebenrett, H.-J., Kozielski, P.-M., Hegner, K., & Welcker, I. (2001). *Lagebild "Jugend heute"* (p. 204). Strausberg: Sozialwissenschaftliches Institut der Bundeswehr.

EC: Directorate-General for Employment, Social Affairs and Equal Opportunities. (2007). In European Commission (Ed.), *Europe's demographic future: Facts and figures on challenges and opportunities* (pp. 184). Luxembourg: Office for Official Publications of the European Communities.

Edmunds, T. (2006). What *are* armed forces for? The changing military roles in Europe. *International Affairs, 82*(6), 1059–1075.

Eighmey, J. (2006). Why do youth enlist? Identification of underlying themes. *Armed Forces & Society, 32*(2), 307–328.

Eriksson, M., & Wallensteen, P. (2004). Armed conflict, 1989–2003. *Journal of Peace Research, 41*(5), 625–636.

Faris, J. H. (1981). The all-volunteer force: Recruitment from military families. *Armed Forces & Society, 7*(4), 545–559.

Feaver, P. D., Hikotani, T., & Narine, S. (2005). Civilian control and civil-military gaps in the United States, Japan, and China. *Asian Perspective, 29*(1), 223–271.

Federal Ministry of Defense. (2006a). *White paper on German security policy and the future of the Bundeswehr* (pp. 126). Bonn: Federal Ministry of Defence.

Federal Ministry of Defense. (2006b). *Personalinformation 2005, Militärisches Personal* (pp. 82). Bonn: PSZ I, Federal Ministry of Defense.

Federal Ministry of Defense. (2009). Data provided by the Social Services and Central Affairs Directorate at the Federal Ministry of Defense in Bonn.

Federal Ministry of Defense. (2010). *Bericht des Generalinspekteurs der Bundeswehr zum Prüfauftrag aus der Kabinettsklausur vom 7. Juni 2010*, Berlin, 63 pages, p. 25.

Federal Ministry of Defense. (2011). *Verteidigungspolitische Richtlinien: Nationale Interessen wahren – Internationale Verantwortung übernehmen – Sicherheit gemeinsam gestalten* (p. 19). Berlin: Bundesministerium der Verteidigung.

Federal Ministry of Transport and Urban Development. (2009). Raumordnungs-prognose 2025/2050. In Bundesamt für Bauwesen und Raumordnung (Ed.), Bevölkerung, private Haushalte, Erwerbspersonen. Bonn.

Federal Statistical Office. (2006a). *Bevölkerung Deutschlands bis 2050 – 11. koordinierte Bevölkerungsvorausberechnung* (pp. 1–72). Wiesbaden: Statistisches Bundesamt.

Federal Statistical Office. (2006b). Leben in Deutschland. Ergebnisse des Mikrozensus 2005. Tabellenanhang zur Pressebroschüre, 191.

Federal Statistical Office. (2008a, April 03). Gesundheit von Kindern und Jugendlichen. *STATmagazin*, 5p.

Federal Statistical Office. (2008b). *Statistisches Jahrbuch 2008 für die Bundesrepublik Deutschland* (p. 725). Wiesbaden: Statistisches Bundesamt.

Federal Statistical Office. (2009). Bevölkerung Deutschlands bis 2060: 12. Koordinierte Bevölkerungsvorausberechnung. Wiesbaden: Statistisches Bundesamt.

Federal Statistical Office. (2011a). Average number of children per woman. https://www.destatis.de/DE/PresseService/Presse/Pressemitteilungen/2011/08/PD11_301_12641.html. Accessed 01 Dec 2011.

Federal Statistical Office. (2011b). Life table for Germany. https://www.destatis.de/DE/ZahlenFakten/GesellschaftStaat/Bevoelkerung/Sterbefaelle/Tabellen/SterbetafelDeutschland.xls?__blob=publicationFile. Accessed 01 Dec 2011.

Fiebig, R. (2008). Bedrohungswahrnehmung und Sicherheitsempfinden. In T. Bulmahn, R. Fiebig, S. Greif, A. Jonas, W. Sender, & V. Wieninger (Eds.), *Sicherheits- und verteidigungspolitisches*

Meinungsklima in der Bundesrepublik Deutschland. Ergebnisse der Bevölkerungsbefragung 2007 (pp. 15–31). Strausberg: Sozialwissenschaftliches Institut der Bundeswehr.

Fischer, G., & Heilig, G. K. (1997). Population momentum and the demand on land and water resources. *Philosophical Transactions: Biological Sciences, 352*(1356), 869–889.

Fishbein, M., & Ajzen, I. (1975). *Belief, attitude, intention, and behavior: An introduction to theory and research.* Reading: Addison-Wesley.

Fleckenstein, B. (2000). Germany: Forerunner of a postnational military? In C. C. Moskos, J. A. Williams, & D. R. Segal (Eds.), *The postmodern military: Armed forces after the cold war* (pp. 80–100). New York/Oxford: Oxford University Press.

Forster, A. (2006). *Armed forces and society in Europe.* New York: Palgrave Macmillan.

Förster, M., & Pearson, M. (2002). Income distribution and poverty in the OECD area: Trends and driving forces. *OECD Economic Studies, 34*(I), 7–39.

Frederking, B. (2003). Constructing post-cold war collective security. *American Political Science Review, 97*(03), 363–378.

Freedman, L. (1991). Demographic change and strategic studies. In L. Freedman & J. Saunders (Eds.), *Population change and European security* (Vol. 1, pp. 7–20). London: Brassey's.

Frieyro de Lara, B. (2010). The professionalization process of the Spanish armed forces. In T. Szvircsev Tresch & C. Leuprecht (Eds.), *Europe without soldiers? Recruitment and retention across the armed forces of Europe* (Queen's policy studies series, pp. 181–193). Montreal/Kingston: McGill-Queen's University Press.

Frisch, T. (2011). *Attraktivitätsmaßnahmen für Freiwilligenstreitkräfte in europäischen Staaten: Sachstand* (p. 15). Berlin: Deutscher Bundestag, Wissenschaftliche Dienste.

Fuller, G. E. (2004). The youth crisis in Middle Eastern society. *Brief paper* (pp. 13). Clinton: Institute for Social Policy and Understanding.

Geisler, E., & Kreyenfeld, M. (2009). *Against all odds: Fathers' use of parental leave in Germany* (Working Paper WP-2009-010, pp. 43). Rostock: Max Planck Institute for Demographic Research.

Gensicke, T. (2002). Individualität und Sicherheit in neuer Synthese? Wertorientierungen gesellschaftliche Aktivität. In D. Shell (Ed.), *Jugend 2002. 14. Shell Jugendstudie* (pp. 139–212). Frankfurt/Main: Fischer.

Gensicke, T. (2006). Zeitgeist und Wertorientierungen. In D. Shell (Ed.), *Jugend 2006. 15. Shell Jugendstudie* (pp. 169–239). Frankfurt/Main: Fischer.

George, M. V., Smith, S. K., Swanson, D. A., & Tayman, J. (2004). Population projections. In D. A. Swanson & J. S. Siegel (Eds.), *Methods and materials of demography* (pp. 561–601). San Diego: Elsevier/Academic Press.

German Bundestag (2008). Umsetzung der Wehrpflicht im Jahr 2007. Antwort der Bundesregierung auf die kleine Anfrage der Abgeordneten Paul Schäfer (Köln), Monika Knoche, Dr. Hakki Keskin, weiterer abgeordneter und der Fraktion Die Linke. Deutscher Bundestag, Drucksache 16/8319 (pp. 32).

German Bundestag (2011). Frauen in der Bundeswehr (pp. 52), Drucksache 17/5664.

Giannakouris, K. (2010). Regional population projections europop2008: Most EU regions face older population profile in 2030. *Statistics in focus* (pp. 20). Brussels: Eurostat.

Gjonça, A., Brockmann, H., & Maier, H. (2000). Old-age mortality in Germany prior to and after reunification. *Demographic Research, 3*(1), 1–29.

Goldstone, J. (2001). Demography, environment, and security: An overview. In M. Weiner & S. S. Russell (Eds.), *Demography and national security* (p. 342). New York/Oxford: Berghahn Books.

Goujon, A., Skirbekk, V., Fliegenschnee, K., & Strzelecki, P. (2007). New times, old beliefs: Projecting the future size of religions in Austria. *Vienna Yearbook of Population Research, 2007*, 237–270.

Gray, C. S. (2007). *War, peace and international relations. An introduction to strategic history.* Abingdon: Routledge.

Grossman, M. (1972). On the concept of health capital and the demand for health. *Journal of Political Economy, 80*, 223–255.

Haas, M. L. (2007). A geriatric peace? The future of U.S. power in a world of aging populations. *International Security, 32*(1), 112–147.

Haas, M. L. (2012). America's golden years? U.S. security in an aging world. In J. A. Goldstone, E. P. Kaufmann, & M. D. Toft (Eds.), *Political demography. How population changes are reshaping international security and national politics* (Vol. 342, pp. 49–62). New York: Oxford University Press.

Haftendorn, H. (2000). Water and international conflict. *Third World Quarterly, 21*(1), 51–68.

Haltiner, K. W. (1998). The definite end of the mass army in Western Europe? *Armed Forces & Society, 25*(1), 7–36.

Haltiner, K. W., & Klein, P. (2005). The European post-cold war military reforms and their impact on civil-military relations. In F. Kernic, P. Klein, & K. Haltiner (Eds.), *The European armed forces in transition: A comparative analysis* (pp. 9–30). Frankfurt am Main: Europäischer Verlag der Wissenschaften.

Haltiner, K., & Kümmel, G. (2009). The hybrid soldier: Identity changes in the military. In G. Kümmel, G. Caforio, & C. Dandeker (Eds.), *Armed forces, soldiers and civil-military relations. Essays in honor of Jürgen Kuhlmann* (pp. 75–82). Wiesbaden: VS Verlag.

Hamil-Luker, J. (2001). The prospects of age war: Inequality between (and within) age groups. *Social Science Research, 30*, 386–400.

Handel, M. (1981). Numbers do count: The question of quality versus quantity. *Journal of Strategic Studies, 4*(3), 225–260.

Harbom, L., & Wallensteen, P. (2009). Patterns of major armed conflicts, 1999–2008. In Stockholm International Peace Research Institute (Ed.), *SIPRI yearbook 2009. Armaments, disarmament and international security* (p. 624). Oxford: Oxford University Press.

Hartley, K. (2006). The British experience with an all-volunteer force. In C. Gilroy & C. Williams (Eds.), *Service to country: Personnel policy and the transformation of western militaries* (pp. 287–312). Cambridge, MA: The MIT Press.

Held, D., McGrew, A., Goldblatt, D., & Perraton, J. (1999). *Global transformations: Politics, economics, and culture*. Stanford: Stanford University Press.

Hesketh, T., & Xing, Z. W. (2006). Abnormal sex ratios in human populations: Causes and consequences. *PNAS, 103*, 13271–13275.

Hofstede, G. (1984). Cultural dimensions in management and planning. *Asia Pacific Journal of Management, 1*(2), 81–99.

Höhn, C., Mai, R., & Micheel, F. (2007). Demographic change in Germany. In I. Hamm, H. Seitz, & M. Werding (Eds.), *Demographic change in Germany. The economic and fiscal consequences* (pp. 9–33). Berlin/Heidelberg: Springer.

Holladay, S. J., & Coombs, W. T. (2004). The political power of seniors. In J. F. Nussbaum & J. Coupland (Eds.), *Handbook of communication and aging research* (pp. 383–406). Mahwah: Lawrence Erlbaum Associates.

Holsti, K. J. (1999). The coming chaos? Armed conflict in the world's periphery. In T. V. Paul & J. A. Hall (Eds.), *International order and the future of world politics* (pp. 283–310). Cambridge: Cambridge University Press.

Homer-Dixon, T. F. (1991). On the threshold: Environmental changes as causes of acute conflict. *International Security, 16*(2), 76–116.

Homer-Dixon, T. F. (1994). Environmental scarcities and violent conflict: Evidence from cases. *International Security, 19*(1), 5–40.

Horne, D. K. (1987). The impact of soldier quality on army performance. *Armed Forces & Society, 13*(3), 443–455.

Horowitz, S. A., & Sherman, A. (1980). A direct measure of the relationship between human capital and productivity. *The Journal of Human Resources, 15*(1), 67–76.

Hosek, J. R. (2003). The soldier of the 21st century. In S. Johnson, M. Libicki, & G. F. Treverton (Eds.), *New challenges, new tools for defense decision-making* (pp. 181–209). Santa Monica: RAND Corporation.

Hosek, J. R., & Peterson, C. (1985). *Enlistment decisions of young men* (p. 83). Santa Monica: RAND Corporation.

Hosek, J. R., & Peterson, C. (1986). *Educational expectations and enlistment decisions* (p. 54). Santa Monica: RAND Corporation.

Hosmer, D. W., & Lemeshow, S. (2000). *Applied logistic regression*. Hoboken: Wiley-Interscience Publication.

Hoyt, T. D. (2003). Security and conflict in the developing world. In M. E. Brown (Ed.), *Grave new world: Security challenges in the 21st century* (pp. 213–229). Washington, DC: Georgetown University Press.

Hradil, S. (2004). *Die Sozialstruktur Deutschlands im internationalen Vergleich*. Wiesbaden: VS Verlag für Sozialwissenschaften.

Hudson, V. M., & Den Boer, A. M. (2002). A surplus of men, a deficit of peace. Security and sex ratios in Asia's largest states. *International Security, 26*(4), 5–38.

Hudson, V. M., & Den Boer, A. M. (2004). *Bare branches: The security implications of Asia's surplus male population* (BCSIA studies in international security). Cambridge, MA: MIT Press.

Hullen, G. (2004). Bevölkerungsentwicklung in Deutschland. Die Bevölkerung schrumpft, altert und wird heterogener. In B. Frevel (Ed.), *Herausforderung demografischer Wandel* (pp. 15–25). Wiesbaden: VS Verlag.

Hundley, R. O. (1999). *Past revolutions, future transformations: What can the history of revolutions in military affairs tell us about transforming the U.S. military?* (p. 98). Santa Monica: National Defense Institute, RAND Corporation.

Huntington, S. (1975). The United States. In M. Crozier, S. Huntington, & J. Watanuki (Eds.), *The crisis of democracy* (pp. 59–118). New York: New York University Press. Total pages of the book: 227.

Huntington, S. P. (1998). *The clash of civilizations and the remaking of world order*. New York: Simon & Schuster.

Hurrelmann, K., Albert, M., Quenzel, G., & Langness, A. (2006). Eine pragmatische Generation unter Druck: Einführung in die Shell-Studie 2006. In S. Deutschland (Ed.), *Jugend 2006 – eine pragmatische Generation unter Druck* (pp. 31–48). Hamburg/Frankfurt am Main: Fischer.

IISS. (2008). Europe. In International Institute for Strategic Studies (Ed.), *The military balance* (Vol. 108:1, pp. 101–204). London: Routledge.

Inglehart, R. (1971). The silent revolution in Europe: Intergenerational change in post-industrial societies. *The American Political Science Review, 65*(4), 991–1017.

Inglehart, R. (1977). *The silent revolution: Changing values and political styles among western publics*. Princeton: Princeton University Press.

Inglehart, R. (1997). *Modernization and postmodernization: Cultural, economic and political change in 43 countries*. Princeton: Princeton University Press.

INKAR. (2007). *Indikatoren, Karten und Graphiken zur Raum- und Stadtentwicklung in Deutschland und in Europa (CD-Rom)*. Bonn: Selbstverlag des Bundesinstituts für Bau-, Stadt- und Raumforschung (BBSR).

Institute for Economics & Peace. (2010). *Global peace index*, New York/Sydney/Oxford. Online at: http://economicsandpeace.org/research/iep-indices-data/global-peace-index

IOM. (2008a). In International Organization for Migration, & Social Science Research Council (Eds.), *Migration and development within and across borders: Research and policy perspectives on internal and international migration* (pp. 373). Geneva/New York: Hammersmith Press. http://publications.iom.int/bookstore/index.php?main_page=product_info&products_id=65

IOM. (2008b). World migration 2008. Managing labor mobility in the evolving global economy. In International Organization for Migration (Ed.), *IOM world migration report series* (Vol. 4, pp. 562). Geneva: IOM.

IOM. (2009). *Combating tobacco use in military and veteran populations*. Washington, DC: Institute of Medicine, The National Academies Press.

Jackson, R., & Howe, N. (2008). *The graying of the great powers – Demography and geopolitics in the 21st century* (p. 184). Washington, DC: Center for Strategic and International Studies, Global Aging Initiative.

JOE. (2008). In Joint Operating Environment (Ed.), *Challenges and implications for the future joint force* (pp. 56). Norfolk: United States Joint Forces Command, Center for Joint Futures.

Johansen, D. (2007). Chapter 2G – recruiting and retention of military personnel: United Kingdom. *Recruiting and Retention of Military Personnel, Final Report of Research Task Group HFM-107, RTO* (Tech. Rep. TR-HFM-107, pp. 2G-1/2G-6). Brussels: North Atlantic Treaty Organization (NATO) & Research Technology Organisation (RTO).

Jonas, A. (2008). Sicherheits- und verteidigungspolitische Einstellungen im Vergleich: Deutschland, Frankreich, Großbritannien, USA. In T. Bulmahn, R. Fiebig, S. Greif, A. Jonas, W. Sender, & V. Wieninger (Eds.), *Sicherheits- und verteidigungspolitisches Meinungsklima in der Bundesrepublik Deutschland. Ergebnisse der Bevölkerungsbefragung 2007* (pp. 157–170). Strausberg: Sozialwissenschaftliches Institut der Bundeswehr.

Junor, L. I., & Oi, J. S. (1996). *A new approach to modeling ship readiness* (p. 71). Alexandria: CNA Corporation, Center for Naval Analyses.

Kagan, F. W. (2006). The U.S. military's manpower crisis [response]. *Foreign Affairs, 85*(4), 97–110.

Kaldor, M. (2007). *New and old wars* (2nd ed.). Stanford: Stanford University Press.

Kane, T. (2006). Who are the recruits? The demographic characteristics of U.S. military enlistment, 2003–2005. *Center for Data Analysis* (Report #06–09, pp. 18). Washington, DC: The Heritage Foundation.

Kaplan, R. D. (1994). The coming anarchy. *Atlantic Monthly, 273*(2), 44–76.

Kapstein, E. B. (1992). *The political economy of national security: A global perspective.* Columbia: University of South Carolina Press.

Kaufmann, E. (2008, March 25–29). *Eurabia? The foreign policy implications of West Europe's religious composition in 2025 and beyond.* Paper presented at the Paper prepared for the International Studies Association (ISA) Annual Conference, San Francisco.

Kent, M. M., & Haub, C. (2005). Global demographic divide. *Population Bulletin, 60*(4), 24.

Kernic, F., Callaghan, J., & Manigart, P. (2002). *Public opinion on European security and defense: A survey of European trends and public attitudes toward CFSP and ESDP.* Vienna: Peter Lang Verlag.

Kilburn, M. R., & Klerman, J. A. (2000). *Enlistment decisions in the 1990s: Evidence from individual-level data* (p. 100). Santa Monica: RAND Corporation.

Klages, H., & Gensicke, T. (2005). Wertewandel und Big-Five-Dimensionen. In S. Schumann (Ed.), *Persönlichkeit. Eine vergessene Größe der empirischen Sozialforschung* (pp. 279–299). Wiesbaden: VS Verlag für Sozialwissenschaften.

Klesges, R. C., Haddock, C. K., Chang, C. F., Talcott, G. W., & Lando, H. A. (2001). The association of smoking and the cost of military training. *Tobacco Control, 10*(1), 43–47.

Kleykamp, M. A. (2006). College, jobs, or the military? Enlistment during a time of war. *Social Science Quarterly, 87*(2), 19.

Klieme, E., Neubrand, M., & Lüdtke, O. (2001). Mathematische Grundbildung: Textkonzeption und Ergebnisse. In J. Baumert, E. Klieme, M. Neubrand, M. Prenzel, U. Schiefele, W. Schneider, et al. (Eds.), *Pisa 2000: Basiskompetenzen von Schülerinnen und Schülern im internationalen Vergleich* (pp. 139–190). Opladen: Leske + Budrich.

KMK. (2007). Vorausberechnung der Schüler- und Absolventenzahlen 2005 bis 2020. In Sekretariat der Ständigen Konferenz der Kultusminister der Länder in der Bundesrepublik Deutschland (Ed.), *Statistische Veröffentlichungen der Kultusministerkonferenz* (pp. 216). Bonn: Sekretariat der Ständigen Konferenz der Kultusminister der Länder in der Bundesrepublik Deutschland. Available online at: http://www.kmk.org/fileadmin/veroeffentlichungen_beschluesse/2007/2007_05_01-Vorausberechnung-Schueler-Absolventen-05-2020.pdf

Köhler, H. (2005, October 10). *Einsatz für Freiheit und Sicherheit.* Speech at the Kommandeurtagung der Bundeswehr, Bonn, 11.

Kohr, H.-U. (1996). *Wertewandel und Soziomoral: Auswirkungen auf die Bundeswehr* (SOWI-Arbeitspapier Nr. 95, pp. 1–19). Strausberg: Sozialwissenschaftliches Institut der Bundeswehr.

Kolodziej, E. A. (2005). *Security and international relations: Themes in international relations.* Cambridge: Cambridge University Press.

Korb, L. J., & Duggan, S. E. (2007). An all-volunteer army? Recruitment and its problems. *Political Science & Politics, 40*(3), 467–471.

Korb, L. J., Ogden, P., & Kagan, F. W. (2006). Jets or GIs? How best to address the military's manpower shortage. *Foreign Affairs, 85*(6), 153–156.

Krause, J. (2006, May 23). *Sicherheitspolitische Herausforderungen.* Paper presented at the conference "Internationale Risiken – sicherheitspolitische und rüstungswirtschaftliche Konsequenzen", Rendsburg.

Krebs, R. R., & Levy, J. S. (2001). Demographic change and the sources of international conflict. In M. Weiner & S. S. Russell (Eds.), *Demography and national security* (pp. 62–105). New York/Oxford: Berghahn Books.

Krepinevich, A. F. (1994). Cavalry to computer; the pattern of military revolutions. *The National Interest, 37*(Fall), 30–42.

Kreyenfeld, M., & Konietzka, D. (2008). Bleibt alles anders. Geburten- und Familienentwicklung in Ost- und Westdeutschland. In N. Werz (Ed.), *Demografischer Wandel* (pp. 50–70). Baden-Baden: Nomos.

Kromeyer-Hauschild, K., Zellner, K., Jaeger, U., & Hoyer, H. (1998). Prevalence of overweight and obesity among school children in Jena (Germany). *International Journal of Obesity, 23*(11), 1143–1150.

Krueger, G. P. (2001). *Military psychology: United States. International encyclopedia of the social & behavioral sciences* (pp. 9868–9873). Oxford: Elsevier.

Kujat, H. (2011). Das Ende der Wehrpflicht – Essay. *Aus Politik und Zeitgeschichte (APuZ 48/2011)*, 3–7.

Kümmel, G. (2003). A soldier is a soldier is a soldier!? The military and its soldiers in an era of globalization. In G. Caforio (Ed.), *Handbook of the sociology of the military* (pp. 417–433). New York: Kluwer Academic/Plenum Publishers.

Kümmel, G., & Leonhard, N. (2005). Casualties and civil-military relations: The German polity between learning and indifference. *Armed Forces & Society, 31*(4), 513–536.

Kutz, M. (1998, October1 9). Berufsbilder und politische Orientierung: Zur soziologischen Typologisierung und politischen Entwicklung des Offizierkorps der Bundeswehr. In *Führungsakademie der Bundeswehr/Fachbereich Sozialwissenschaften: Beiträge zu Lehre und Forschung, 6/98*, Hamburg.

Lampert, T., & Thamm, M. (2007). Tabak-, Alkohol- und Drogenkonsum von Jugendlichen in Deutschland. *Bundesgesundheitsbl – Gesundheitsforsch – Gesundheitsschutz, 5/6*, 600–608.

Langness, A., Leven, I., & Hurrelmann, K. (2006). Jugendliche Lebenswelten: Familie, Schule, Freizeit. In *Jugend 2006 – eine pragmatische Generation unter Druck* (pp. 49–102). Frankfurt/Main: Fischer.

Lawrence, G. H., & Legree, P. J. (1996). In Army Research Institute for the Behavioral and Social Sciences (Ed.), Military enlistment propensity: A review of recent literature (pp. 79). Alexandria: Army Research Institute for the Behavioral and Social Sciences.

Leahy, E., Engelman, R., Vogel, C. G., Haddock, S., & Preston, T. (2007). *The shape of things to come: Why age structure matters to a safer, more equitable world* (p. 100). Washington, DC: Population Action International.

Lee, R. D., & Edwards, R. D. (2001). The fiscal impact of population change. In J. S. Little & R. K. Triest (Eds.), *Seismic shifts: The economic impact of demographic change* (pp. 189–219). Boston: Federal Reserve Bank of Boston.

Lescreve, F., & Schreurs, B. (2007). Chapter 2A – Recruiting and retention of military personnel: Belgium. *Recruiting and retention of military personnel, Final Report of Research Task Group HFM-107, RTO* (Tech. Rep. TR-HFM-107, pp. 2A-1/2A-9). Brussels: North Atlantic Treaty Organization (NATO) & Research Technology Organisation (RTO).

Lesthaeghe, R. (1983). A century of demographic and cultural change in Western Europe: An exploration of underlying dimensions. *Population and Development Review, 9*(3), 411–435.

Lesthaeghe, R., & Surkyn, J. (1988). Cultural dynamics and economic theories of fertility change. *Population and Development Review, 14*(1), 1–45.

Leuprecht, C. (2006). Die unvollendete Revolution: Post-nationale Streitkräfte. In U. vom Hagen (Ed.), *Armee in der Demokratie: Zum Verhältnis von zivilen und militärischen Prinzipien* (pp. 31–49). Wiesbaden: VS Verlag.

Leuprecht, C. (2010). Socially representative armed forces: A demographic imperative. In T. Szvircsev Tresch & C. Leuprecht (Eds.), *Europe without soldiers? Recruitment and retention across the armed forces of Europe* (Queen's policy studies series, pp. 35–54). Montreal/Kingston: McGill-Queen's University Press.

Levy, J. S. (1988). Domestic politics and war. *Journal of Interdisciplinary History, 18*(4), 653–673.

Leyk, D., Rohde, U., Gorges, W., Ridder, D., Wunderlich, M., Dinklage, C., et al. (2006). Physical performance, body weight and BMI of young adults in Germany 2000–2004: Results of the physical-fitness-test study. *International Journal of Sports Medicine, 27*, 642–647.

Leyk, D., Rohde, U., Gorges, W., Wunderlich, M., Rüther, T., Wamser, P., et al. (2007). Erste Ergebnisse der Studie "Fit-fürs-Leben": Übergewicht und Bewegungsmangel bei Heranwachsenden und jungen Erwachsenen *Wehrmedizinische Monatsschrift, 51*(5-6/2007), 143–147.

Lippert, E. (1992). *Was ist, was kann und was soll Militärsoziologie?* München: Sozialwissenschaftliches Institut der Bundeswehr.

Lipsey, R. E. (2007). Home- and host-country effects of foreign direct investment. In R. E. Baldwin & L. A. Winters (Eds.), *Challenges to globalization: Analyzing the economics* (pp. 333–379). Chicago: University of Chicago Press.

Longman, P. (2004). *The empty cradle. How falling birthrates threaten world prosperity and what to do about it.* Cambridge, MA: Basic Books.

Luttwak, E. N. (1994). Where are the great powers? At home with the kids. *Foreign Affairs, 73*(4), 23–28.

Luttwak, E. N. (1995). Toward post-heroic warfare. *Foreign Affairs, 74*(3), 109–122.

Luttwak, E. N. (2003). *Strategie. Die Logik von Krieg und Frieden* (400pp.). Lüneburg: Zu Klampen Verlag.

Mackinlay, J. (1994). Improving multifunctional forces. *Survival, 36*(3), 149–173.

Mai, R. (2008). Demographic change in Germany. *European View, 7*, 287–296.

Maizière, T. d. (2011). Abgabe einer Regierungserklärung durch den Bundesminister der Verteidigung zur Neuausrichtung der Bundeswehr. *Plenarprotokoll 17/112* (pp. 84). Berlin: Deutscher Bundestag.

Manigart, P. (2003). Restructuring of the armed forces. In G. Caforio (Ed.), *Handbook of the sociology of the military* (pp. 323–343). New York: Kluwer Academic/Plenum Publishers.

Manigart, P. (2005). Risks and recruitment in postmodern armed forces: The case of Belgium. *Armed Forces & Society, 31*(4), 559–582.

Manton, K. G., Stallard, E., & Corder, L. (1997). Changes in the age dependence of mortality and disability: Cohort and other determinants. *Demography, 34*(1), 135–157.

Mare, R. D., & Winship, C. (1984). The paradox of lessening racial inequality and joblessness among black youth: Enrollment, enlistment, and employment, 1964–1981. *American Sociological Review, 49*, 39–55.

Mason, D., & Dandeker, C. (2009). Evolving UK policy on diversity in the armed services: Multiculturalism and its discontents. *Commonwealth and Comparative Politics, 47*(4), 393–410.

McGuire, M. C. (2000). Concepts of defense economics for the 21st century. *Defence and Peace Economics, 11*(1), 17–30.

McIntosh, C. A. (1983). *Population policy in Western Europe – Responses to low fertility in France, Sweden and West Germany.* New York: M.E. Sharpe, Inc.

McKie, R. (2004, January 25). Living with Britain's population time bomb. *The Observer.*

McLaughlin, R., & Wittert, G. (2009). The obesity epidemic: Implications for recruitment and retention of defense force personnel. *Obesity Reviews, 10*(6), 693–699.

McMichael, A. J., McKee, M., Shkolnikov, V., & Valkonen, T. (2004). Mortality trends and setbacks: Global convergence or divergence? *Lancet, 363*, 1155–1159.

Menard, S. (1995). *Applied logistic regression analysis* (Quantitative applications in the social sciences series). Thousand Oaks: Sage.

Milkoreit, M. (2007, Fall). Taking the civil dimension of security seriously: NATO as the post-conflict reconstruction organization. *NATO Review*. Also accessible here: http://www. nato.int/docu/review/2007/Military_civilian_divide/post-conflict_reconstruction_organization/ EN/index.htm

Miller, R. A. (1995). Domestic structures and the divisionary use of force. *American Journal of Political Science, 39*(3), 760–785.

Mintz, A., & Huang, C. (1990). Defense expenditures, economic growth, and the 'peace dividend'. *The American Political Science Review, 84*(4), 1283–1293.

Moelker, R., Olsthoorn, P., Bos-Bakx, M., & Soeters, J. (2005). *From conscription to expeditionary armed forces: Trends in the professionalisation of the Royal Netherlands armed forces* (p. 60). Breda: Royal Netherlands Military Academy, Faculty of Military Sciences.

Moniz, D. (2005, February 20). Military offering more, and bigger, bonuses. *USA Today*. http:// usatoday30.usatoday.com/news/washington/2005-02-20-military-bonuses_x.htm

Moravcsik, A. (1997). Taking preferences seriously: A liberal theory of international politics. *International Organization, 51*(4), 513–553.

Moskos, C. C. (1977). From institution to occupation: Trends in military organization. *Armed Forces & Society, 4*(1), 41–50.

Münz, R. (2007). *Aging and demographic change in European societies: Main trends and alternative policy options'* (SP Discussion Paper No. 0703). Washington, DC: Social Protection Advisory Service/The World Bank.

Murasko, J. E. (2009). Socioeconomic status, height, and obesity in children. *Economics and Human Biology, 7*(3), 376–386.

Murray, M. P., & McDonald, L. L. (1996). *Recent recruiting trends and their implications for models of enlistment supply* (p. 83). Santa Monica: RAND Corporation.

National Report on Education. (2008). *Education in Germany 2008: An indicator-based report including an analysis of transitions subsequent to lower secondary education*. Bielefeld: Bertelsmann Verlag.

NATO. (2007). Recruiting and retention of military personnel. In North Atlantic Treaty Organisation (Ed.), *Final report of research task group* (p. 516). Brussels: Research and Technology Organisation (RTO) of NATO.

Nichiporuk, B. (2000). *The security dynamics of demographic factors* (p. 74). Santa Monica: RAND Corporation.

Nickel, J., Ravens-Sieberer, U., Richter, M., & Settertobulte, W. (2008). Gesundheitsrelevantes Verhalten und soziale Ungleichheit bei Kindern und Jugendlichen. In M. Richter, K. Hurrelmann, A. Klocke, W. Melzer, & U. Ravens-Sieberer (Eds.), *Gesundheit, Ungleichheit und jugendliche Lebenswelten* (pp. 63–92). Weinheim/München: Juventa.

Nieva, V. F., Wilson, M. J., Norris, D. G., Greenlees, J. B., & Laurence, J. (1997). *Enlistment intentions and behaviors: Youth and parental models* (p. 260). Alexandria: U.S. Army Research Institute for the Behavioral and Social Sciences.

NRC (2003). *Attitudes, aptitudes, and aspirations of American youth: Implications for military recruitment* (National Research Council, Committee on the youth population and military recruitment). Washington, DC: National Academies Press.

NRC (Ed.). (2006). *Assessing fitness for military enlistment. Physical, medical, and mental health standards.* (National Research Council, Committee on the youth population and military recruitment). Washington, DC: National Academies Press.

Nyce, S. A., & Schieber, S. J. (2005). *The economic implications of aging societies: The costs of living happily ever after*. Cambridge: Cambridge University Press.

OECD. (2005). Water and violent conflict. In Development Assistance Committee (Ed.), *Issues brief* (pp. 1–10). Washington, DC: OECD.

OECD. (2008). *Education at a glance* (p. 521). Paris: Centre for Educational Research and Innovation; Organisation for Economic Co-Operation and Development.

OECD. (2009). *Education at a glance* (p. 475). Paris: Centre for Educational Research and Innovation; Organisation for Economic Co-Operation and Development.

O'Hare, W., & Bishop, B. (2006). *U.S. rural soldiers account for a disproportionately high share of casualties in Iraq and Afghanistan* (Fact sheet no. 3, The Carsey Institute Reports on Rural America, p. 2). Durham: University of New Hampshire.

Opper, E., Worth, A., Wagner, M., & Bös, K. (2007). Motorik-Modul im Rahmen des Kinder- und Jugendgesundheitssurveys: Motorische Leistungsfähigkeit und körperlich-sportliche Aktivität von Kindern und Jugendlichen in Deutschland. *Bundesgesundheitsbl – Gesundheitsforsch – Gesundheitsschutz, 50*, 879–888.

Organski, A., Bueno de Mesquita, B., & Lamborn, A. (1972). The effective population in international politics. In R. L. Clinton & R. Kenneth Godwin (Eds.), *Political science in population studies*. Lexington: Heath.

Orvis, B. R., & Asch, B. J. (2001). *Military recruiting: Trends, outlook and implications* (p. 57). Santa Monica: RAND Corporation.

Orvis, B. R., Childress, M., & Polich, J. M. (1992). *Effect of personnel quality on the performance of patriot air defense system operators* (p. 88). Santa Monica: RAND Corporation.

Orvis, B. R., Sastry, N., & McDonald, L. L. (1996). *Military recruiting outlook: Recent trends in enlistment propensity and conversion of potential enlisted supply* (p. 80). Santa Monica: RAND Corporation.

Postel, S. L., & Wolf, A. T. (2001). Dehydrating conflict. *Foreign Policy*, No. 126 (September/October), 60–67. http://links.jstor.org/sici?sici=0015-7228%28200109%2F10%290%3A126%3C60%3ADC%3E2.0.CO%3B2-M

Poterba, J. M. (1997). Demographic structure and the political economy of public education. *Journal of Policy Analysis and Management, 16*(1), 48–66, Special Issue: Serrano V, Priest: 25th Anniversary.

Pötzsch, O. (2007). In Statistisches Bundesamt (Ed.), *Geburten in Deutschland* (pp. 34). Wiesbaden: Statistisches Bundesamt. https://www.destatis.de/DE/Publikationen/Thematisch/Bevoelkerung/Bevoelkerungsbewegung/BroschuereGeburtenDeutschland0120007079004.pdf?__blob=publicationFile

Puente, J., & Blanco, S. (2007). Chapter 2E – Recruiting and retention of military personnel: Spain. *Recruiting and retention of military personnel, Final Report of Research Task Group HFM-107, RTO* (Tech. Rep. TR-HFM-107, pp. 2E-1/2E-13). Brussels: North Atlantic Treaty Organization (NATO) & Research Technology Organisation (RTO).

Ratha, D., & Shaw, W. (2007). *South-south migration and remittances* (p. 38). Washington, DC: Development Prospects Group, The World Bank.

Ravens-Sieberer, U., Thomas, C., & Erhart, M. (2003). Körperliche, psychische und soziale Gesundheit von Jugendlichen. In K. Hurrelmann, A. Klocke, W. Melzer, & U. Ravens-Sieberer (Eds.), *Jugendgesundheitssurvey: Internationale Vergleichsstudie im Auftrag der Weltgesundheitsorganisation* (pp. 19–98). Weinheim/München: Juventa.

Reinberg, A., & Hummel, M. (2003). Bildungspolitik: Steuert Deutschland langfristig auf einen Fachkräftemangel zu? *IAB Kurzbericht Nr. 9/2003* (pp. 1–7). Nürnberg: Institut für Arbeitsmarkt- und Berufsforschung.

Richardson, R. (2010). Recruitment and retention of ethnic cultural minorities in the Dutch armed forces. In T. Szvircsev Tresch & C. Leuprecht (Eds.), *Europe without soldiers? Recruitment and retention across the armed forces of Europe* (Queen's policy studies series, pp. 21–34). Montreal/Kingston: McGill-Queen's University Press.

Richter, M., & Settertobulte, W. (2003). Gesundheits- und Freizeitverhalten von Jugendlichen. In K. Hurrelmann, A. Klocke, W. Melzer, & U. Ravens-Sieberer (Eds.), *Jugendgesundheitssurvey: Internationale Vergleichsstudie im Auftrag der Weltgesundheitsorganisation* (pp. 99–157). Weinheim/München: Juventa.

Ridge, M., & Smith, R. (1991). UK military manpower and substitutability. *Defence Economics, 2*(4), 283–294.

Rijsberman, F. R. (2006). Water scarcity: Fact or fiction? *Agricultural Water Management, 80*(1–3), 5–22.

Riley, J. (2001). *Rising life expectancy: A global history*. Cambridge: University Press.

Risse-Kappen, T. (1991, July). Public opinion, domestic structure, and foreign policy in liberal democracies. *World Politics, 43*(4), 479–512.

Rohde, U., Erley, O., Rüther, T., Wunderlich, M., & Leyk, D. (2007). Leistungsanforderungen bei typischen soldatischen Einsatzbelastungen. *Wehrmed Mschr, 51*(5–6), 138–142.

Rose, E. (2007). Siblings and soldiers: Family background and military service in the all-volunteer era. Working paper (pp. 36). Seattle: Department of Economics, University of Washington.

Rosen, S. (1986). The theory of equalizing differences. In O. Ashenfelter & R. Layard (Eds.), *Handbook of labor economics* (Vol. I, pp. 641–692). Amsterdam: North Holland.

Rosenau, J. N. (1994). Armed force and armed forces in a turbulent world. In J. Burk (Ed.), *The military in new times. Adapting armed forces to a turbulent world* (pp. 25–61). Boulder/San Francisco/Oxford: Westview Press.

Russett, B. M. (1971). An empirical typology of international military alliances. *Midwest Journal of Political Science, 5*(2), 262–289.

Ryder, N. B. (1965). The cohort as a concept in the study of social change. *American Sociological Review, 30*(6), 843–861.

Sandell, R. (2006). Coping with demography in NATO Europe: Military recruitment in times of population decline. In C. L. Gilroy & C. Williams (Eds.), *Service to country – personnel policy and the transformation of western militaries* (BCSIA studies in international security, pp. 65–96). Cambridge, MA: The MIT Press.

Sandler, T., & Hartley, K. (1995). *The economics of defense* (Vol. 3). Cambridge: Cambridge University Press.

Sarkesian, S. C. (1985). Low-intensity conflict: Concepts, principles and policy guidelines. *Air University Review, 36*(2), 4–23.

Schaffer, H. I. (1992). *Lebenskonzepte und Zeiterfahrungen junger Männer: Zur Bedeutung gewandelter Lebensvorstellungen für die Bundeswehr* (pp. 1–55). München: Sozialwissenschaftliches Institut der Bundeswehr.

Schjølset, A. (2010). NATO and the women: Exploring the gender gap in the armed forces. *PRIO Paper* (pp. 56). Oslo: Peace Research Institute Oslo (PRIO).

Schmid, T. (2011). The world from our living room: No joy in their political sovereignty, no interest in foreign policy. *IP Journal*. Online publication, date published: 17 February 2011. https://ip-journal.dgap.org/en/ip-journal/regions/world-our-living-room

Schreiner, K. H. (2005, October 19). *Welchen Soldaten braucht die deutsche Bundeswehr? – Normative Erwartungen.* 50. Gesamtkonferenz der hauptamtlichen katholischen Militärgeistlichen und Pastoralreferenten, Speech by the brigadier general of Fü S I, Potsdam, 20.

Sciubba, J. D. (2011). *The future faces of war: Population and national security.* Santa Barbara: Praeger.

Scribner, B. L., Smith, D. A., Baldwin, R. H., & Phillips, R. L. (1986). Are smart tankers better? AFQT and military productivity. *Armed Forces & Society, 12*(2), 193–206.

Segal, M. W. (1986). The military and the family as greedy institutions. *Armed Forces & Society, 13*(9), 9–38.

Segal, D., & Segal, M. W. (2004). America's military population. *Population Bulletin, 59*(4), 44.

Seifert, R. (1992). *Soldatische Subjektivität, gesellschaftlicher Wandel und Führungsanforderungen: Plädoyer für eine Subjektperspektive in der Militärsoziologie.* München: Sozialwissenschaftliches Institut der Bundeswehr.

Seitz, H. (2007). The impact of demographic change on fiscal policy. In I. Hamm, H. Seitz, & M. Werding (Eds.), *Demographic change in Germany. The economic and fiscal consequences* (pp. 129–163). Berlin/Heidelberg: Springer.

Seitz, H., & Kempkes, G. (2007). Fiscal federalism and demography. *Public Finance Review, 35*(3), 385–413.

Shahla, H., Fischer, A., & Hubert, T. (2005). *Mikrozensus scientific use file 2003, Dokumentation und Datenaufbereitung, Zuma-Methodenbericht 2005/06* (pp. 20). Mannheim: ZUMA, German Microdata Lab (GML).

Shaw, M. (1991). *Post-military society: Militarism, demilitarization and war at the end of the twentieth century.* Philadelphia: Temple University Press.

Smith, H. (2005). What costs will democracies bear? A review of popular theories of casualty aversion. *Armed Forces & Society, 31*(4), 487–512.

Smith, R. (2006). *The utility of force: The art of war in the modern world.* London: Penguin.

Smith, D. A., Sylwester, S. D., & Villa, C. M. (1991). Army reenlistment models. In C. L. Gilroy, D. K. Horne, & D. A. Smith (Eds.), *Military compensation and personnel retention: Models and evidence* (pp. 43–173). Washington, DC: United States Army Research Institute for the Behavioral and Social Sciences.

Smith, S. K., Tayman, J., & Swanson, D. A. (2001). *State and local population projections: Methodology and analysis* (The plenum series on demographic methods and population analysis). New York: Kluwer Academic/Plenum Publishers.

Snow, D. M. (1999). *The shape of the future: World politics in a new century* (3rd ed.). New York: M.E. Sharpe.

Sprout, H., & Sprout, M. (1968). The dilemma of rising demands and insufficient resources. *World Politics, 20*(4), 660–693.

Stepanova, E. (2008). Trends in armed conflicts. In Stockholm International Peace Research Institute (Ed.), *Sipri yearbook 2008. Armaments, disarmament and international security* (p. 640). Oxford: Oxford University Press.

Streeck, W. (2007). Politik in einer alternden Gesellschaft: Vom Generationenvertrag zum Generationenkonflikt? In P. Gruss (Ed.), *Die Zukunft des Alterns: Die Antwort der Wissenschaft* (p. 331). München: C.H. Beck Verlag.

Strengmann-Kuhn, W. (1999). Armutsanalysen mit dem Mikrozensus? In P. Lüttinger (Ed.), *Zuma-Nachrichten Spezial* (Vol. 6, pp. 376–402). Mannheim: GESIS.

Strohmeier, G. (2007). *Bericht zur Mitgliederbefragung des deutschen Bundeswehr-Verbandes. Umfrage zur Berufszufriedenheit* (p. 170). Passau: Deutscher Bundeswehrverband.

Sumer, H. C. (2007). Chapter 3J – Individual differences and later turnover. In North Atlantic Treaty Organization (NATO) (Ed.), *Recruiting and Retention of Military Personnel, Final Report of Research Task Group HFM-107, RTO* (Tech. Rep. TR-HFM-107, pp. 3J-1/3J-16). Brussels: North Atlantic Treaty Organization (NATO) & Research Technology Organisation (RTO).

Syed, F., & Morrow, R. (2007). Recruiting and retention of military personnel: Canada. *Recruiting and retention of military personnel, Final Report of Research Task Group HFM-107, RTO* (Tech. Rep. TR-HFM-107, pp. 2B-1/2B-14). Brussels: North Atlantic Treaty Organization (NATO) & Research Technology Organisation (RTO).

Szvircsev Tresch, T. (2005). *Europas Streitkräfte im Wandel: Von der Wehrpflichtarmee zur Freiwilligenstreitkraft. Eine empirische Untersuchung europäischer Streitkräfte 1975–2003.* Zürich: Universität Zürich.

Szvircsev Tresch, T. (2010). Recruitment of military professionals by European all-volunteer forces as exemplified by Belgium, the Netherlands, and Slovenia. In T. Szvircsev Tresch & C. Leuprecht (Eds.), *Europe without soldiers? Recruitment and retention across the armed forces of Europe* (Queen's policy studies series, pp. 145–164). Montreal/Kingston: McGill-Queen's University Press.

Teachman, J. D., Call, V. R. A., & Segal, M. W. (1993). The selectivity of military enlistment. *Journal of Political and Military Sociology, 21*(2), 287–309.

Teitelbaum, M. S., & Winter, J. M. (1985). *Fear of population decline.* London: Academic.

Tellis, A. J., Bially, J., Layne, C., & McPherson, M. (2003). *Measuring national power in the postindustrial age* (p. 212). Santa Monica: RAND Corporation.

The World Bank. (2011). *World development indicators*, Washington, D.C. Online at http://data.worldbank.org/indicator

Thomas, G. W. (1984). Military parental effects and career orientation under the AVF: Enlisted personnel. *Armed Forces & Society, 10*(2), 293–310.

Toset, H. P. W., Gleditsch, N. P., & Hegre, H. (2000). Shared rivers and interstate conflict. *Political Geography, 19*(8), 971–996.

U.S. Department of Defense. (2001). *Quadrennial defense review report* (p. 79). Washington, DC: Department of Defense.

U.S. Department of Defense. (2008). *National defense strategy* (p. 29). Washington, DC: Department of Defense.

UN. (2004). *World urbanization prospects. The 2003 revision. Highlights* (pp. 335). New York: United Nations, Department of Economic and Social Affairs, Population Division.

UN. (2006). *World migrant stock: The 2005 revision population database.* New York: United Nations, Department of Economic and Social Affairs, Population Division. http://esa.un.org/migration/. Accessed 01 Nov 2009.

UN. (2007). *World population prospects: The 2006 revision. Highlights* (pp. 114). New York: United Nations, Department of Economic and Social Affairs, Population Division.

UN. (2008). *World urbanization prospects. The 2007 revision. Highlights* (pp. 244). New York: United Nations, Department of Economic and Social Affairs, Population Division.

UN. (2010). *World population prospects: The 2010 revision.* New York: United Nations, Department of Economic and Social Affairs, Population Division. Online at http://esa.un.org/wpp/

UN. (2011). *World population prospects: The 2010 revision population database.* New York: United Nations, Department of Economic and Social Affairs, Population Division. http://esa.un.org/unpd/wpp/unpp/panel_indicators.htm. Accessed 15 Dec 2011.

UNCTAD. (2008). *World investment report 2008: Transnational corporations, and the infrastructure challenge* (p. 411). New York/Geneva: United Nations Conference on Trade and Development.

UNEP. (1999). *Global environment outlook 2000.* Nairobi/Kenya: Division of Environmental Information, Assessment and Early Warning, United Nations Environment Programme.

UNFPA. (2007). In Deutsche Stiftung Weltbevölkerung (Eds.), *Weltbevölkerungsbericht 2007. Urbanisierung als Chance: Das Potenzial wachsender Städte nutzen* (pp. 120). Hannover/Stuttgart: Deutsche Stiftung Weltbevölkerung.

Urdal, H. (2004). The devil in the demographics: The effect of youth bulges on domestic armed conflict, 1950–2000. *Social Development Papers, Conflict Prevention & Reconstruction.* Washington, DC: World Bank.

Urdal, H. (2006). A clash of generations? Youth bulges and political violence. *International Studies Quarterly, 50*(3), 607–629(623).

Urdal, H. (2011, July 21–22). A clash of generations? Youth bulges and political violence. *United Nations Expert Group Meeting on Adolescents, Youth and Development.* New York: Population Division, Department of Economic and Social Affairs, United Nations Secretariat.

van Bladel, J. (2004, December). *European public security management practice.* Paper presented at the Joint workshop "Security Sector Reform" and "Regional Stability Track", Budapest.

van Crefeld, M. (2008). *Changing face of war: Combat from the Marne to Iraq.* New York: Presidio Press.

van Creveld, M. (1991). *The transformation of war.* New York: The Free Press.

van de Kaa, D. J. (1987). Europe's second demographic transition. *Population Bulletin, 42*(1), 1–58.

van de Kaa, D. J. (1997). Options and sequences: Europe's demographic patterns. *Journal of Population Research, 14*(1), 1–29.

van de Ven, C., & Bergman, R. (2007). Chapter 2D – Recruiting and retention of military personnel: The Netherlands. *Recruiting and retention of military personnel, Final Report of Research Task Group HFM-107, RTO* (Tech Rep. TR-HFM-107, pp. 2D-1/2D-13). Brussels: North Atlantic Treaty Organization (NATO) & Research Technology Organisation (RTO).

van der Meulen, J., & Soeters, J. (2005). Considering casualties: Risk and loss during peacekeeping and warmaking. *Armed Forces & Society, 31*(4), 483–486.

van der Meulen, J., & Soeters, J. (2007). Introduction. In J. Soeters & J. v. d. Meulen (Eds.), *Cultural diversity in the armed forces. An international comparison* (pp. 1–14). Oxon: Routledge.

Vaupel, J. W., & von Kistowski, K. G. (2008). Die neue Demografie und ihre Implikationen für Gesellschaft und Politik. In N. Werz (Ed.), *Demografischer Wandel* (pp. 33–49). Baden-Baden: Nomos.

Vincent, J. A. (2003a, September 4–6). *Demography, politics and old age*, British Society for Gerontology Annual Conference, Newcastle upon Tyne, p. 1.

Vincent, J. A. (2003b, September 23–26). *New forms of global political economy and ageing societies*. Paper presented at the European Sociological Association Conference "Ageing in Europe: Challenges of Globalisation for Ageing Societies", Murcia, 24.

Vincent, J. A. (2005). Understanding generations: Political economy and culture in an ageing society. *The British Journal of Sociology, 56*(4), 579–599.

von Bredow, W. (2007). Conceptual insecurity: News wars, MOOTW, CRO, terrorism, and the military. In G. Caforio (Ed.), *Social sciences and the military* (pp. 163–180). New York: Routledge.

von Bredow, W., & Kümmel, G. (1999). *Das Militär und die Herausforderung globaler Sicherheit. Der Spagat zwischen traditionalen und nicht-traditionalen Rollen* (SOWI-Arbeitspapier Nr. 119, p. 29). Strausberg: Sozialwissenschaftliches Institut der Bundeswehr.

Vroom, V. H. (1964). *Work and motivation*. New York: Wiley.

Wabitsch, M. (2004). Kinder und Jugendliche mit Adipositas in deutschland. *Bundesgesundheitsbl – Gesundheitsforsch – Gesundheitsschutz, 3*, 251–255.

Wabitsch, M., Kunze, D., Keller, E., Kiess, W., & Kromeyer-Hauschild, K. (2002). Adipositas bei Kindern und Jugendlichen in Deutschland: Deutliche und anhaltende Zunahme der Prävalenz – Aufruf zum Handeln. *Fortschritte der Medizin/Originalien, 120*(4), 99–106.

Wagschal, U., Metz, T., & Schwank, N. (2008). Ein 'demografischer Frieden'? Der Einfluss von Bevölkerungsfaktoren auf inner- und zwischenstaatliche Konflikte. *Zeitschrift für Politikwissenschaft, 18. Jahrgang*(3), 353–383.

Waltz, K. (1979). *Theory of international politics*. Reading: Addison Wesley.

Ware, V. (2010). Whiteness in the glare of war: Soldiers, migrants and citizenship. *Ethnicities, 10*(3), 313–330.

Warner, J. T., & Asch, B. J. (1995). The economics of military manpower. In K. Hartley & T. Sandler (Eds.), *Handbook of defense economics* (Vol. 1, pp. 347–398). New York: Elsevier.

Warner, J. T., & Asch, B. J. (2000). Themes in defence manpower economics and challenges for the future. *Defence and Peace Economics, 11*(1), 93–103.

Warner, J. T., Simon, C. J., & Payne, D. M. (2001). *Enlistment supply in the 1990s: A study of the navy college fund and other enlistment incentive programs* (p. 114). Arlington: Defense Manpower Data Center.

Weiner, M., & Russell, S. S. (2001). *Demography and national security*. New York/Oxford: Berghahn Books.

Weiner, M., & Teitelbaum, M. (2001). *Political demography, demographic engineering*. New York/Oxford: Berghahn Books.

Williams, C. (2004). Introduction. In C. Williams (Ed.), *Filling the ranks: Transforming the U.S. military personnel system* (Belfer Center for Science and International Affairs, pp. 1–28). Cambridge, MA: John F. Kennedy School of Government, Harvard University.

Williams, C., & Gilroy, C. (2006). The transformation of personnel policies. *Defence Studies, 6*(1), 97–121.

Williams, C., & Seibert, B. (2011). *Von der Wehrpflichtigen- zur Freiwilligenarmee. Erkenntnisse aus verbündeten Staaten* (p. 28). Cambridge, MA: Weatherhead Center for International Affairs.

Wilson, C. (2001). On the scale of global demographic convergence 1950–2000. *Population and Development Review, 27*(1), 155–171.

Wilson, M. J., Gay, N. L., Allen, B. F., & Celeste, J. F. (1988). *The army enlistment decision: A selected, annotated bibliography* (p. 113). Rockville: U.S. Army Research Institute for the Behavioral and Social Sciences.

Winkler, J. D. (1999). Are smart communicators better? Soldier aptitude and team performance. *Military Psychology, 11*(4), 405–422.

Wöhlcke, M. (1996). *Sicherheitsrisiken aus Umweltveränderungen*. Ebenhausen: Stiftung Wissenschaft und Politik.

Wöhlcke, M., Höhn, C., & Schmid, S. (2004). *Demographische Entwicklungen in und um Europa – politische Konsequenzen* (Band 69, Aktuelle Materialien zur internationalen Politik). Baden-Baden: Nomos.

Wolf, A. T. (1999). Water and human security. *Aviso: An Information Bulletin on Global Environmental Change and Human Security, 3*, 1–7.

Wolf, T. (2009, July 26). Rentengarantie: Buhlen um die Senioren. *FOCUS.* http://www.focus.de/finanzen/altersvorsorge/rente/tid-14972/rentengarantie-buhlen-um-die-senioren_aid_419948.html

Wolffsohn, M. (2009, August 21). Die Bundeswehr ist eine Unterschichtenarmee. *WELT Online.* http://www.welt.de/politik/deutschland/article4368744/Die-Bundeswehr-ist-eine-Unterschichtenarmee.html

Woodruff, T., Kelty, R., & Segal, D. R. (2006). Propensity to serve and motivation to enlist among American combat soldiers. *Armed Forces & Society, 32*(3), 353–366.

Ziegler, U., & Doblhammer, G. (2005). Steigende Lebenserwartung geht mit besserer Gesundheit einher: Risiko der Pflegebedürftigkeit sinkt. *Demografische Forschung Aus Erster Hand, Jahrgang 2*(Nr. 1), 1–2.

Index

W. Apt, *Germany's New Security Demographics: Military Recruitment
in the Era of Population Aging*, Demographic Research Monographs,
DOI 10.1007/978-94-007-6964-9, © Springer Science+Business Media Dordrecht 2014

Printed by Printforce, the Netherlands